Industry 4.0 Topical Conference 2018

Topical Conference at the 2018 AIChE Spring Meeting and 14th Global Congress on Process Safety

Orlando, Florida, USA
22 - 25 April 2018

ISBN: 978-1-5108-6433-7

Printed from e-media with permission by:

Curran Associates, Inc.
57 Morehouse Lane
Red Hook, NY 12571

Some format issues inherent in the e-media version may also appear in this print version.

Printed by Curran Associates, Inc. (2018)

For permission requests, please contact AIChE
at the address below.

AIChE
120 Wall Street, FL 23
New York, NY 10005-4020

Phone: (800) 242-4363
Fax: (203) 775-5177

www.aiche.org

Additional copies of this publication are available from:

Curran Associates, Inc.
57 Morehouse Lane
Red Hook, NY 12571 USA
Phone: 845-758-0400
Fax: 845-758-2633
Email: curran@proceedings.com
Web: www.proceedings.com

TABLE OF CONTENTS

Some abstracts in this book refer to images that were not made available by the publisher

513988 Digitalization Strategies in an Age of Rapid Change

Abstract:

As with many large chemical producers, Dow has launched a wide ranging strategy towards achieving significant value through implementation of developing technologies that will speed decision making, make work safer, and improve reliability. A common thread in all these initiatives is the unification of IT and analytics into the digitalization of the work place. Digitalization, in its most classic form, simply means turning information into a computer-readable format that can be easily rendered in a variety of formats. A digitalization strategy, therefore, means planning for the use of that information from inception, through storage, to end-user, all sitting upon an appropriate digital thread. In manufacturing in general, applications are as limitless as humans can find practical use, but in the chemical industry there are particular areas that Dow has identified as key for extracting the value we seek. We will discuss Dowâ€™s broad initiatives in two areas: Business / Supply Chain, and Manufacturing / Engineering, as well as featuring the formation of two key capabilities that will support this strategy â€“ Dowâ€™s Chicago-based Digital Marketplace supporting our commercial strategy, and the newly opened Houston-based Digital Operations Center supporting our M&E strategy.

513481 Digital Twins for Predicting Early Onset of Failures Flow Valves

Abstract:

Unplanned shutdowns in petrochemical refineries due to equipment failures cause significant financial losses. Across our operations, Shell has been collecting real-time data for decades. At present, more than 10 million operational variables per minute are collected, streamed, archived, integrated with operational control systems. There is enormous potential to exploit this data further. With predictive analytics and machine learning algorithms, unexpected failures and unnecessary maintenance can be avoided, saving assets millions of dollars per annum in optimized maintenance and deferment avoidance.

This paper presents the results of two different proof-of-concept study carried out in Shell Pernis and Shell Martinez manufacturing sites, to use unlabelled historical process control data to develop a digital twin algorithm which predicts failures. Experiments for the use case were performed on multiple control valves. The target was to verify, whether the machine learning methods are capable of distinguishing between normally behaving valves from abnormally behaving valves. The difference between the prediction and the measured values, i.e. the error rate, should be as low as possible for the normal valves while being as large as possible for the abnormal valves.

Multiple solutions based on artificial neural networks and statistical approaches were implemented to model the normal behavior of the monitored systems in Pernis. Mismatches with predictions of the modelled systems were then used to predict failures in advance, where the prediction horizon reached more than a month for some use cases. 4-layer GRUs (Gated Recurrent Unit) with tanh activation functions and an input sequence length of 4 samples provided the best results. GRUs were 7% faster to train than LSTMs (Long Short-Term Memory) while reducing the prediction error by 15%. Furthermore, the approach allows for predicting failure with high accuracy, characterising failing signals with deviations that are 5x larger than the deviations during normal operation. This indicates that machine learning models can predict failures in petrochemical refineries for the studied use case, without the need for industry-specific knowledge, if the model is trained with data representing fault- free operation.

In case of Martinez, first, a traditional mass balance with meters upstream and downstream of the target flow element was utilized. A secondary accuracy check occurred by developing a novel, machine learning based predictive deterioration model that relied on first principles and statistical multivariate regression to augment and validate traditional mass balancing. This model successfully verified that the mass balance approach provided acceptable meter accuracies in this instance and allowed engineers to track predicted flow meter performance vs. actual metering. Going forward, these models will enable equipment deterioration analysis and be the catalytic step change towards predictive maintenance.

511696 Accelerating Time to Value By Integrating Smart Technology and Business Operations

Abstract:

As smart manufacturing technologies have proliferated in factories, business operations have struggled to keep pace with the new requirements these technologies present. With advanced manufacturing systems come demands on sourcing, maintenance, scheduling, information management, human resource and financial functions. Automation offers higher levels of throughput and process consistency. But it can come at a cost. Raw materials may need to be delivered to tighter tolerances. Information on pedigree may need to be provided to smart analytics. Reduced operator touch time might be replaced by increased technical skill to troubleshoot equipment or monitor quality through real time charting. Few suppliers of advanced technologies are equipped to understand these critical business demands, let alone incorporate new requirements into their designs. The result is a protracted period of debug and redesign after equipment is delivered and accepted by factories, characterized by higher scrap rates, increase labor cost and missed schedules. This presentation will offer a different approach to developing advanced manufacturing systems that achieves rapid return on investment by including business outcomes in the development. Examples will be provided that show how integration of business operations and system design achieved better performance in less time.

513432 Challenges for Big Data Use of Production Process Data

Abstract:

There are a number of initiatives for the future of manufacturing such as Industry 4.0(1), Digital Transformation Initiative(2), and Smart Manufacturing Leadership Coalition(3). These bring together a variety of concepts to create a seamless connection between the management of various data sources, advanced computer simulations/analysis, and adaptive user/machine/organization interactions. All of these include process data as part of their corporate integration, but there are a number of issues with this type of data that will need to be address if it is going to be made transparent to the user. In particular, this is the incorporation of knowledge within Big Data(4). This presentation will look at some of these issues and consider examples where these issues impact the understanding of the data.

In particular, the presentation will consider issues of the compression of production data, examples about various types of errors in the measurements, the time delays within a process, and the high correlation of process operation. The knowledge applied to these issues includes the details about the processes that ranges from the physical information about the equipment to the materials and conditions used to the theoretical understanding of the system. The knowledge needed also includes the type of sensors that are used for the measurements and the management and sampling of the data.

In addition, there is a need to understand the various intended use of the data and the appropriate analysis for that intention. The computations should be different if the intent is to predict future action compared to an intent of understanding why a certain defect occurred. This is a difference that is often ignored when analyzing data.

The examples are derived from actual production data to provide actual insight to how these issues have been traditional handled and provide some examples of how advanced analytics has helped with these issues. The intention is to encourage further development of working with production data that deals with these issues and that could be truly fulfill some of the vision of data transparency and advance analytics for manufacturingâ€™s next evolution.

1. Kagermann H, Wahlster W, Helbig J. Recommendations for implementing the strategic initiative INDUSTRIE 4.0. Final report of the Industrie 4.0 Working Group. Acatech- National Academy of Science and Engineering; 2013.
2. World_Economic_Forum. Digital Transformation Initiative Chemistry and Advanced Materials Industry. Geneva, Switzerland: World Economic Forum; 2017.
3. Davis J, Edgar T, Porter J, Bernaden J, Sarli M. Smart manufacturing, manufacturing intelligence and demand-dynamic performance. Computers & Chemical Engineering. 2012;47(Supplement C):145-56.
4. Wuest T, Weimer D, Irgens C, Thoben K-D. Machine learning in manufacturing: advantages, challenges, and applications. Production & Manufacturing Research. 2016;4(1):23-45.

513906 Risk Analysis for Count Data

Abstract:

The methodology for performing risk analysis calculation in the presence of process and measurement variation has been well established in the literature [1-4]. However current methods are limited to continuous distribution, specifically normal distribution. Even though a majority of quality measurements made within Dow are continuous in nature, there are still discrete measurements such as counts of defects, grade level assignments to name a few. In addition to this, the advances made in the field of metrology have also allowed higher resolution measurements be made on the process or products. One such capability is the ability to count low-level defects continuously on product, and such count data typically do not fit a normal distribution. The work here is to extend the risk analysis on low level defect count data which can be best described using the Poisson distribution.

[1] Burdick, R. K.; Borror, C. M.; and Montgomery, D. C. (2003). A Review of Methods for Measurement Systems Capability Analysis. *Journal of Quality Technology*, 35 (4), pp. 342â€"354.

[2] Mader, D. P.; Prins, J.; and Lampe, R. E. (1999). The Economic Impact of Measurement Error. *Quality Engineering* 11(4), pp. 563â€"574.

[3] Mottonen, Matti; Belt, Pekka; Harkonen, Janne; Haapasalo, Harri; Kess, Pekka; Manufacturing Process Capability and Specification Limits, *The Open Industrial and Manufacturing Engineering Journal*, 2008, 1, 29-36.

[4] Basnet, Chuda; Case, Kenneth E. (1992). A Review of Methods for Measurement Systems Capability Analysis. *Quality Engineering*, 4 (3), pp. 383â €"397.

513401 Dealing with Small Data in Biopharmaceutical Batch Process Monitoring: A Machine-Learning Approach

Abstract:

In recent years, biotechnology-based products have been gaining increasing visibility and success in treating chronic diseases, such as arthritis, diabetes and cancer. As per the report [1] published in 2005, approximately 27% of the new medicines in active development are in fact biotechnology-based products. Another report published in 2013 by the United States Food and Drug Administration (US FDA) states that a growing percentage, now over 40%, of all pharmaceutical industry research are now biopharmaceutical rather than classical drugs.

To efficiently monitor and control biopharmaceutical processes, multivariate statistical techniques are commonly deployed for batch process monitoring (BPM). A BPM framework uses multivariate statistical models (e.g., principal component analysis (PCA) and partial least squares (PLS)) to capture common-cause variations in the batch [2]. The control charts (e.g., Hotelling T2 and the squared prediction error (SPE) statistics) and control limits are then used to determine whether a new batch demonstrates normal operating behavior or not. An alarm is raised if a batch is statistically different from the normal operations.

Despite over two decades of research in BPM, the biopharmaceutical BPM framework suffers from a unique challenge the Low-N problem (or small-data problem). The Low-N problem represents a scenario where a product has a limited production history, denoted here by N. It is common for companies to have only one or two runs for a new drug product at a manufacturing facility. While a new product may require a limited number of runs to meet clinical or early commercial demand, it also creates a Low-N scenario for the product. In terms of BPM, a Low-N scenario poses several challenges. First, under the Low-N, it is nontrivial to capture the common-cause variations in its entirety. Second, the predictive capabilities of PCA and PLS are less accurate under the Low-N scenario. Further, under the Low-N scenario, model over-fitting becomes much harder to avoid and the effects of outliers are much more pronounced.

The Low-N problem is a longstanding, industry-wide problem in biopharmaceutical manufacturing that challenges the theoretical foundations and practical applicability of the existing BPM platform. We propose an approach to transition from a Low-N scenario to a Large-N scenario by generating an arbitrarily large number of in silico batch data sets. The proposed method is a combination of hardware exploitation and algorithm development. To this effect, we propose a block-learning method for a Bayesian non-parametric model of a batch process, and then use probabilistic programming to generate an arbitrarily large number of dynamic in silico campaign data sets. The proposed solution not only alleviates the monitoring issues associated with a Low-N scenario, it is also compatible with the industrial BPM framework. To the best of authors knowledge, this is the first method that describes a systematic approach to address the small data problem using the tools for big data. The efficacy of the proposed solution is elucidated on an industrial biopharmaceutical process.

References

[1]. Hamilton G. "The Biotechnology Market Outlook: Growth Opportunities and Effective Strategies for Licensing and Collaboration". Dublin: Research and Markets. 2005.

[2]. P. Nomikos and J. F. MacGregor, "Monitoring batch processes using multiway principal component analysis, AIChE Journal, vol. 40, no. 8, pp. 1361–1375, 1994.

513487 Analysing Big Data in Dairy Processing, By Throwing Most of It Away

Abstract:

Dairy processing in New Zealand has a desire to reduce out-of-specification production of value-added milk powders. While the process is tightly controlled, and such events are rare, the detection of some faults takes up to three days post-production, and leads to significant downgrades given very high production rates. Grading tests indicate if powders are fit for release, but are not clear measurements useful for data analysis. It is hard to identify the plant operating conditions that lead to failure, and furthermore, the underlying physical mechanisms are not always well understood.

The examination of multiple industrial plants spread over geographically separate sites, all producing similar products requires careful management of significant amounts of data. Ideally this "Big Data" approach would underpin models describing the process, which could be used to predict key functional properties of the milk powder, and deliver early warning signals before product was to slip out of specification. Extensive work was done in the very difficult task of collecting and aligning several years of process and quality data, from plants around New Zealand, representing a range of designs, ages, geographical locations, process control schemes, and data storage types.

However, as recent publications have shown, and this study reinforced, the creation of the dataset is typically a much larger job than the actual modelling and analysis, which at times can seem as trivial as sending the data to a regression function. It is becoming increasingly evident that advanced technical knowledge is not simply waiting to be unlocked by the "Big Data Revolution". Our early approaches led to a range of models that suffered from poor predictability and were of little practical use. It appeared to be a combination of several factors, where measurements to discern the key physical phenomena were not available (or possible), and that there was not enough data in the main regions of interest, due to the rarity of failure.

Whilst adding new instruments is difficult and expensive, substantial improvements in predictability were made, not by collecting even more data, but by "throwing much of it away". Changes in plant operations, equipment, and products mean that only `small data' models can be applied to subsets of the data. It was necessary to go back to first principles and apply detailed operational knowledge to find ways to improve the time alignment of process data and quality samples. Statistical resampling of data sets was applied to give more weight to rare occurrences, as well as using categorical modelling to find these occurrences.

This approach, and the need for fundamental engineering information and careful examination of the physical process was reinforced by investigation of a different big data set for cream cheese production. The data was analysed to give an overview of the behaviour of the plant, but progress in modelling and prediction required significant work and low-level calculations to dissect the data. Rather than multivariate data analysis, this required actual engineering knowledge to understand how the plant is automatically controlled, how the operators physically conducted manual tasks, and the vagaries of how the different instruments measure and are recorded.

This paper will describe two examples of how "Big Data" - that is the collection and preparation of the data - is not a solution to our problems, rather than a tool which still needs careful investigation and engineering knowledge to understand engineering problems. It is the unfortunate situation that often the biggest data still does not contain enough information, either because key information is not measured, or because there is not enough data in the regions we are really interested in. In this case, models that are intended to predict certain failures need to have substantial data that exhibits these failures. We describe this problem and how we were able to address it.

514288 Low-Touch Machine Learning Is Fulfilling the Promise of APM (poster)

Abstract:

New methods and cutting-edge technologies are driving asset performance management (APM) well beyond historical capabilities, rapidly increasing its bottom-line value. Technologies such as cloud computing, data science and machine learning are now being integrated with automated methodologies directly into APM solutions.

This wave of integration firmly places advanced analytical techniques into the hands of operators and engineers with previously unimagined scale and ease of use. The incremental progress in APM over the last 20 years pales in comparison to what's now possible through digital transformation.

Low-touch machine learning is the key catalyst to scale APM's potential well beyond existing first principles-based solutions and "armies" of consultant engineers and data scientists. A widespread integration of machine learning in APM will mark a transition from estimated engineering and statistical models towards measuring actual patterns of asset behavior.

Manufacturing facilities staff can now readily extract value from decades of existing design and operations data to better manage and optimize asset performance. This "low-touch" machine learning method continuously embraces changes in asset behavior, empowering real-time APM value creation. Vetted and tested across diverse industries, scalable across multiple assets and powered by cloud and parallel computing, low-touch machine learning ushers in a new era of performance and optimization for all industries.

514063 Big Data Analytics Platforms: The Key to Enabling the Analytic Enterprise (poster)

Abstract:

Analytics and the Industrial Internet of Things are the new hot buzzwords these days, but analyzing data is nothing new, and manufacturers have been generating data off the line for decades. So why all the hype? The last few years have seen the evolution and advancement of several technologies that are converging to allow for unprecedented analysis and insights into business problems. But leveraging these takes more than just smart coders. Analytic insights need to be deployed into production and action has to be taken to truly realize the benefits. Find out why analytic platforms are the key to enabling the Analytic Enterprise.

514064 Industry 4.0 and Driving Better Organizational Decisions (poster)

Abstract:

Industry 4.0 is more than just about the Industrial Internet of Things, it is also about leveraging the Industrial Internet of People (IIOP). This talk will focus on using the tools now available to bring sensors, data, and analytics together to drive better organizational understanding and continuous improvement. Who are the key stake holders, and what do they need to improve to improve their contribution to the way a chemical plant is operated? How do you structure these projects to ensure successful deployments? This talk will focus on practical use cases and examples of Industry 4.0 in action.

514067 Early-Stage Digital Transformation, a Case Study (poster)

Abstract:

Enthought transforms the business of science with artificial intelligence. We are a world-class provider of scientific software solutions and digital transformation services to companies in a range of industries, from chemical manufacturing to oil and gas upstream research and production to aerospace, finance, and others. For more than 15 years Enthought has been building artificial intelligence and machine learning based solutions with science and engineering at the core.

In this talk, we present a case study of digital transformation at a polymer design and manufacturing company looking to improve their data management and increase extraction of the value hidden in their data. We will cover details of the on-going project, including steps taken to digitize previously manual, paper-based operations; changes in workflow and processes to support digitalization; and examples of the practical applications of machine learning techniques. In addition, we will discuss lessons learned, successes and pitfalls in the project. Positive transformations were not limited to the business only but also included new opportunities and personal transformation for employees and a positive change of culture at the company.

514070 Integrated Technical Computing for Process Engineering (poster)

Abstract:

Process engineering leverages data, domain knowledge, and modeling to design, scale, and continuously improve plants, from R&D to pilot to production. The workflow fundamentally involves data analysis and visualization, model development, and process optimization.

Data analysis starts with preprocessing of the dataset, to impute missing values, correct errors, and often impart structure. Descriptive analysis is used to characterize what has been observed, for each individual quantity, as well as compute correlations between quantities. Visualization is used to rapidly develop understanding regarding the relationships between predictors and responses.

Insights gained from data analysis are subsequently leveraged to create a model, to describe the responses as mathematical functions of the predictors. In simple cases, first principles understanding is confirmed, and a governing equation can be written. In cases of intermediate complexity, the basic mechanisms for observed behavior are described by a theoretical framework, and parameter estimation or process identification methods can be used. Complex cases are typified by a multiplicity of superposed physical effects and absence of an economic functional form â€"in these cases, machine learning techniques are required to infer input-output relationships.

Once the model has been created, predictions can be made about the responses that would be recorded if an experiment based on the predictor values was conducted. Since the model enables the parameter space to be explored continuously, optimization methods can be employed to find the combination of predictors that minimizes a specified function of the responses. The optimal conditions so determined inform the selection of the plant operating setpoint. Sensitivity studies are conducted to define the plant operating region around the setpoint, and thus specify the control system requirements.

There are numerous software tools available which address select subsets of the process engineering workflow. MATLAB is one of very few that cover it end-to-end, and is distinguished in that it also offers utilities for distributed computing, big data processing, automatic code generation, and real-time interfacing with other software tools â€"including process simulators.

In this talk we describe MATLAB functionality relevant to each stage of the process engineering workflow, and provide application examples and graphics to bring the concepts to life. Specific topics will include data analysis (tall objects, logical indexing, statistical plotting), model development (principal components, system identification, neural networks), and optimization (linear programming, pattern search, and genetic algorithms).

514071 Tony Starck Needs Jarvis to Run Iron Man. What's the Future in AI for Continuous Process Industries? (Poster)

Abstract:

Why bigdata is such a game changer? We have run plants with automation and scada levels for 30 years, and we encounter a lot of data. We visualize, we set, we adjust, we regulate. With more and more advanced technologies. We make a lot of decisions, impacting the performance of our plants. The bigdata era creates another level of opportunity. By crossing all kinds of information, we open the gate of a new worldâ€¦ of complexity. We understand that there is a gold mine awaiting our discovery, but our biological brain, the main technology we use to dig into it, is now outdated. Our brains will not crunch the bigdata! We need new technology : artificial intelligence.

Today, the most advanced continuous process plants are about to become autonomous. Transforming the decisions of process optimization to dedicated AI. This is coming fast for all of us. What does already exist? How do we implement it for what profit? What can we expect for the future? How can this benefit us?

514072 Approaches to Multiblock Modeling and Dimension Reduction for Combining Disparate Data Types (poster)

Abstract:

Continuous manufacturing, which is being strongly promoted by the FDA, is an innovative approach to pharmaceutical processing that utilizes information from multiple sensors throughout the entire process. Due to the numerous factors that need to be monitored and controlled at the same time, a multi-step approach is necessary for optimizing individual unit operations as well as the overall process. In this presentation we will show some approaches to combining process and spectroscopic data, which can be applied to process monitoring for continuous manufacturing as well as for batch processes.

513463 Manufacturing Analytics - the Journey so Far

Abstract:

The goal of Enterprise Manufacturing Intelligence (EMI) is to make these high value relevant data available in real time to all levels of production management (Operations to Plant and Business Management). Therefore, plants can take advantage of this information to make improvements in raw material usage, asset utilization by reducing unnecessary process wait times and customer satisfaction through improved product quality by reducing product and process variation. This applies to both batch and continuous processes.

At The Dow Chemical Company, we have been actively pursuing EMI since 2012. Since that time we have done 70+ EMI implementations in a variety of environments. This poster talk will discuss the different audiences / kinds of problems our installations have sought to address. We will also discuss some challenges (both technical and cultural) and how we overcame them. Finally we will discuss how these experiences have shaped our view of the future for EMI at Dow.

516856 Applying Predictive Analytics to Develop Actionable Demand Forecasts

Abstract:

The Dow Chemical Company is a leading chemical manufacturer in the world. Dow manufactures plastics, chemicals, and agriculture products. With a presence in over 150 countries, it employs over 50,000 people worldwide. Dow serves to customers in different market segments, with a wide variety of products made for each one. The magnitude of Dow makes demand forecasting a challenge. Dow is using hierarchical statistical demand forecasting techniques to accurately forecast demand for the key products and customers. This poster is about the data and predictive analytics methods and tools used to develop the actionable demand forecasts.

513044 A Novel Prediction Method of the Chemical Leakage Range with Artificial Intelligence: Machine Learning for Chemical Species

Abstract:

This study focuses on the prediction of the chemical leakage range by using artificial intelligence (AI). A multi-variable regression model is developed by deep neural network (DNN) method. The DNN model is formulated by Python with Tensorflow library. The model contains 12 input variables, 70 hidden layers, and 3 output variables. Each layer has Rectified Linear Unit (ReLu) function as an activation function and layers are fully connected. All weights are initialized by Xavier initializer. The chemical leakage range data set is generated by the commercial software, ALOHA. The simulation results show the hazardous zones (AEGL-1, AEGL-2, AEGL-3) under various conditions. To train the AI, the data set is configured with variables: chemical species, wind speed (1~7 m/s), temperature (10~35 â„ƒ), ground roughness (open country, open water), stability classes (A~F), and tank storages (200~4,000 kg). Total date sets of each chemical are 30,240, which are divided into 25,920 training data and 4,320 test data. Training data set is the given information to artificial intelligence and the test data set is unknown information. As the results, the developed model performs high accuracy over 95 % on average for all tested chemicals. The result shows that artificial intelligence figure out the regularity of the data on its own and make predictions with high accuracy. The developed model can be extended to predict chemical leakage ranges for unknown chemicals as well as chemical mixtures by using chemical property information as the input data. The same method could also be used for quick estimation of fire and explosion accidents. This kind of research is expected to contribute to real-time response of corresponding organization in the chemical process safety field.

This research was supported by Disaster-Safety Platform Technology Development Program of the National Research Foundation of Korea(NRF) funded by the Ministry of Science and ICT. (NRF-2017M3D7A1085361)

512180 Estimating Physical Properties of the Productsof an Atmospheric Distillation Columnby Support Vector Regression

Abstract:

Atmospheric distillation column is one of the most important units in an oil refinery where crude oil that is little use as it is, is fractioned into its more valuable constituents. Different fractions of the crude distillation unit are then further processed in downstream conversion and purification units to produce final products such as gasoline, diesel and jet fuel. The physical properties and the quantity of the final products vary depending on the physicochemical properties of the crude oil being processed and the operation parameters of the atmospheric distillation column. The process operators should keep the physical properties of hydrocarbon products in specified limits and operate the crude distillation unit according to instructions from the production planning department for maximizing the profit from operations. The physical properties of different crude distillation unit fractions must be measured by taking a sample from the stream periodically and analyzing these samples in a laboratory with appropriate equipment or by online analyzers that are very expensive to install, operate and maintain. Almost all of the state-of- the art online equipment has a time lag to complete the analysis in real time due to complexity of the analyses. Therefore, the laboratory or on-line measurements become available to decision makers and operator with a time lag. The intermittent nature of the measurements leads to sub-optimal control of the unit or in some cases leads to off-spec products that have considerably less economic value than the products that are within specified limits. As a result, estimation of the physical properties from online plant data and implementation of a soft sensor has a potential benefit in improving the profit. Among different approaches, the support vector regression shows great promise in machine learning due to its ability in generalizing well to unseen test data.

We aim to estimate the physical properties of the hydrocarbon products of an atmospheric distillation column using the large volume of data obtained from real-time and online sensors installed in the system. Linear, Polynomial and Gaussian Radial Basis Function (Gaussian RBF) kernels have been tested and the SVR parameters have been optimized by embedding k-fold cross validation using a variety of algorithms including genetic algorithm (GA), grid search (GS) and non-linear programming (NLP). The performance of SVR has been compared against artificial neural network (ANN) method and robust quality estimator (RQE) already functioning in the ADC. The testing results suggest that SVR with any of the integrated kernels performs well in estimating the property and generalizes well to blind test data. Compared to RQE, the mean testing error of estimation has been improved by 31% with SVR from 6.4ËšC to 4.4ËšC and the standard deviation of estimation error is improved by 23% with SVR from 7.3 ËšC to 5.6 ËšC.

513440 Genetic Process Visualization Using Parametric t-Sne

Abstract:

Distributed control systems (DCS) are essential for modern chemical plants. Plant operators use the DCS to monitor the operation in real time and keep the process within safe and productive domains of operation. Commonly, the monitoring interface in the DCS is designed for single variable analysis, which is expressed in two ways: tags with current values on the process diagram or a line chart on a time axis. Based on the thresholds of each variable, alarms notify personnel if any single variable drifts out of its safe limitations. These strategies are clear to operators and are also simple to implement, but they can only provide limited information of the process due to the handicap of single variable statistics. In such a system, the DCS is blind to any correlation information among multiple variables.

In our work, we proposed a generic process visualization method using dimensionality reduction techniques based on deep learning, which is the state-of-the-art technique that uses deep neural networks (DNN) to extract features hidden within large data sets [1]. It has already achieved notable success in many areas, including image analysis [2], natural language processing [3] and even human-level intelligence for control tasks [4]. For the objective of process monitoring and data visualization, we adopted a deep learning model, namely parametric t-SNE [5]. As a dimensionality reduction method derived from t-SNE [6], it uses a DNN to optimize the projection of the data into the latent space by minimizing the Kullback-Leibler divergence from original space. In this method, different measurements in each data record are considered as a whole, allowing correlation features to be learned parametrically by the DNN. The objective of utilizing the parametric t-SNE in process visualization is to project the high-dimensional measurement onto a 2D map, where different normal operating regions and faults can be separated. Thus, a notable change can be observed in the latent space which can provide qualitative information regarding the process operation to the plant operator.

In practice, we observed that the nonuniformed variance in each variable can create bias, causing the DNN to greedily learn the features from only the most contributing variables in the training set. In other words, variables that show less variance in the training set are muted in the model, which is misleading for visualizing unknown conditions of the process in the future. Therefore, beyond following the two-stage training procedure of the parametric t-SNE, we used combinatorial variation creation method to introduce variance equally to each variable. Process data from a pyrolysis reactor was used as a case study for the proposed method. Data from one operating cycle was used to training the DNN and the trained model was tested on another cycle. The visualization model is especially effective in collaboration with data-driven fault detection methods (e.g. PCA-T2), where it can provide more in-depth information about process operation on the 2D map. This can relieve the operators from staring at the numerical values and line charts of each variable in interest and move their work to a more intuitive interface.

Reference

[1] LeCun, Yann, Yoshua Bengio, and Geoffrey Hinton. "Deep learning." *Nature* 521.7553 (2015): 436-444.

[2] Krizhevsky, Alex, Ilya Sutskever, and Geoffrey E. Hinton. "Imagenet classification with deep convolutional neural networks." *Advances in neural information processing systems*. 2012.

[3] Graves, Alex, Abdel-rahman Mohamed, and Geoffrey Hinton. "Speech recognition with deep recurrent neural networks." *Acoustics, speech and signal processing (icassp), 2013 ieee international conference on*. IEEE, 2013.

[4] Mnih, Volodymyr, et al. "Human-level control through deep reinforcement learning." *Nature* 518.7540 (2015): 529-533.

[5] van der Maaten, Laurens. "Learning a parametric embedding by preserving local structure." *RBM* 500.500 (2009): 26.

[6] Maaten, Laurens van der, and Geoffrey Hinton. "Visualizing data using t-SNE." *Journal of Machine Learning Research* 9.Nov (2008): 2579-2605.

514287 Low-Touch Machine Learning Is Fulfilling the Promise of APM

Abstract:

New methods and cutting-edge technologies are driving asset performance management (APM) well beyond historical capabilities, rapidly increasing its bottom-line value. Technologies such as cloud computing, data science and machine learning are now being integrated with automated methodologies directly into APM solutions.

This wave of integration firmly places advanced analytical techniques into the hands of operators and engineers with previously unimagined scale and ease of use. The incremental progress in APM over the last 20 years pales in comparison to what's now possible through digital transformation.

Low-touch machine learning is the key catalyst to scale APM's potential well beyond existing first principles-based solutions and "armies" of consultant engineers and data scientists. A widespread integration of machine learning in APM will mark a transition from estimated engineering and statistical models towards measuring actual patterns of asset behavior.

Manufacturing facilities staff can now readily extract value from decades of existing design and operations data to better manage and optimize asset performance. This "low-touch" machine learning method continuously embraces changes in asset behavior, empowering real-time APM value creation. Vetted and tested across diverse industries, scalable across multiple assets and powered by cloud and parallel computing, low-touch machine learning ushers in a new era of performance and optimization for all industries.

514056 Big Data Analytics Platforms: The Key to Enabling the Analytic Enterprise

Abstract:

Analytics and the Industrial Internet of Things are the new hot buzzwords these days, but analyzing data is nothing new, and manufacturers have been generating data off the line for decades. So why all the hype? The last few years have seen the evolution and advancement of several technologies that are converging to allow for unprecedented analysis and insights into business problems. But leveraging these takes more than just smart coders. Analytic insights need to be deployed into production and action has to be taken to truly realize the benefits. Find out why analytic platforms are the key to enabling the Analytic Enterprise.

513880 Industry 4.0 and Driving Better Organizational Decisions

Abstract:

Industry 4.0 is more than just about the Industrial Internet of Things, it is also about leveraging the Industrial Internet of People (IIOP). This talk will focus on using the tools now available to bring sensors, data, and analytics together to drive better organizational understanding and continuous improvement. Who are the key stake holders, and what do they need to improve to improve their contribution to the way a chemical plant is operated? How do you structure these projects to ensure successful deployments? This talk will focus on practical use cases and examples of Industry 4.0 in action.

509732 Digitalization in the Process Industries – How and Where to Implement

Abstract:

Digitalization... It's THE buzz word in the industry. But what is it? And how can you go about implementing digitalization in your plant and achieve the benefits that it promises? The goals of the presentation are to provide attendees with an overview of digitalization, including the expected benefits, and explain some practical examples of how it can be implemented. Specific focus will be on the vast array of field devices in your plant that include an increasing amount of intelligence and methods you can use to capitalize on the information they provide from the following perspectives... engineering, operations, and maintenance. Also discussed will be ways to use digital fieldbus technology to provide effective communications between smart field devices and the process automation system.

Specific questions to be answered include...

- *What is digitalization and what are the benefits?*
- *What information is available and how can it be effectively transmitted?*
- *How can I use this information to more productively operate and maintain my plant?*
- *How can this technology be used to more efficiently engineer my automation system?*

514096 Operational Analytics: Transforming Operational Data into Actionable Insights

Abstract:

The advent of digital revolution has now enabled Chemical Industry with numerous tools that could be leveraged to transform our operational data into actionable insights. Key opportunities with digitization include better visualization, integrated planning and execution for value-chain optimization, that results in smarter production, intelligent response to changes in raw materials, process and equipment conditions, reduce energy and waste along with prevention of asset breakdown, safety and environmental issues.

The goal is to link processing plant performance to operational insights, integrating traditionally siloed data and KPIâ€™s for a more holistic approach to optimizing Supply Chain Plant Operations. Integrating operational targets with actual operational insights. The operational events become the new transactions in a process plant.

This paper discusses strategic data hierarchy methodology, findings, and challenges in the ongoing journey of implementing an analytics platform that evolves from being retrospectively descriptive to anticipatively prescriptive.

511688 Empirical Models for Analyzing BIG Data. What's the Difference?

Abstract:

Many issues around "BIG Data" (eg. collection, warehousing, integration, cloud services) are just infrastructure issues that need to be improved in order to ultimately use the data to extract actionable information. If the data analysis step is not performed well, then most of the infrastructure effort will have been wasted. To analyze historical data, one needs to make use of models, usually empirical – such as regression, data mining (deep learning neural networks, decision trees, etc.) or latent variable models. My PhD supervisor in Statistics (G.E.P. Box) used to often say "**All models are wrong, but some are useful**". The problem is that most people lump all empirical models into one category – "empirical models" - as though any of these models are interchangeable irrespective of the nature of the data or the objectives of the problem.

But, whether a model is useful depends upon three factors:

1. The objectives of the model
2. The nature of the data used for the modeling
3. The regression method used to build the model

From an objective point of view, there are basically two major classes of models – those to be used for **passive** use and those to be used for **active** use. Models for **passive** use are intended to be used just to passively observe the process in the future. Such passive applications include classification, inferentials or soft sensors, and process monitoring (MSPC). For such **passive** uses one does not need or even want causal models, rather one wants to just model the normal variations common to the operating process. Historical data is ideal for building such models. Models for **active** use are intended to be used to actively alter the process. Such active applications include using the models to optimize or control the process or to trouble-shoot process problems or gain causal information from the data. For such active use one needs **causal** models. Causality implies that for any active changes in the adjustable or manipulatable variables in the process, the model will reliably predict the changes in the output of interest.

The problem is that most of the data available in industry is historical operating data and such data almost never contain causal information on individual variables. This poses a major problem if we want causal models. The question then becomes what analysis methods are useful for obtaining **active** models from historical operating data?

Machine Learning (ML) methods are currently the rage in "BIG Data" communities. These include deep learning neural networks, and massive decision trees. These new ML approaches are improvements on the older "shallow" neural networks (a few layers connecting all variables to all nodes) and single large decision trees, both of which led to overfitting of the data and to large variances in the predictions. The newer ML approaches are aimed at overcoming some of these deficiencies. Deep learning NN's use many simplified layers, regularization and averaging to reduce the effective number of parameters and the overfitting. New decision trees involve building many decision trees based on fewer randomly selected variable and then averaging or voting on the results to effectively reduce the variance and bias of the results.

These newer ML methods can be very good for passive uses. But they cannot be used for extracting interpretable or causal models from historical data for active use. With historical data, there are an infinite number of models that can arise from any of these machine learning methods, all of which might provide good predictions of the outputs, but none of which is unique or causal. This does not allow for meaningful interpretations, even more so if the results come from averaging or voting on many (eg often 1000 or more) models.

Nevertheless, these ML models have proven to be very powerful in **passive** applications – eg. deep learning NN's for image analysis, and Random Forests for medical diagnosis, both of which are passive applications (where there is no interest in altering the image or the patient).

Latent Variable models such as PLS (Partial Least Squares or Projection to Latent Variables) were developed specifically to handle "BIG Data" where the real number of things affecting the process is much smaller than the number of measured variables. Typically, the number of latent variables needed to extract useable information from hundreds of process variables is more in the order of 3 to 10, implying that the true number of degrees of freedom that affect the process is often quite small.

These latent variable models such as PLS or PCR (Principal Component Regression) are a total break from the classic statistical regression of machine learning models in that they assume that the input or regressor (X) space and the output (Y) space are not of full statistical rank and so they provide models for both the X and Y spaces rather than just the Y space. Without simultaneous models for both the X and Y spaces, there can be no model uniqueness nor any model interpretabilty or causality when using historical data. LV models do provide causality, but only in the reduced dimension space of the latent variables as this is the only space within which the historical process has varied. By moving the latent variables one can reliably predict the outputs (Y) – implying causality for the latent variables. However, to move the LV's one cannot just adjust individual x variables, rather combinations of the X variables that define the LV's as defined by the X space model.

ProSensus has used this uniqueness and causality provided by LV models for many years to trouble-shoot and understand processes and to optimize both processes and products based on actionable information extracted from "BIG" historical data.

513504 Early-Stage Digital Transformation, a Case Study

Abstract:

Enthought transforms the business of science with artificial intelligence. We are a world-class provider of scientific software solutions and digital transformation services to companies in a range of industries, from chemical manufacturing to oil and gas upstream research and production to aerospace, finance, and others. For more than 15 years Enthought has been building artificial intelligence and machine learning based solutions with science and engineering at the core.

In this talk, we present a case study of digital transformation at a polymer design and manufacturing company looking to improve their data management and increase extraction of the value hidden in their data. We will cover details of the on-going project, including steps taken to digitize previously manual, paper-based operations; changes in workflow and processes to support digitalization; and examples of the practical applications of machine learning techniques. In addition, we will discuss lessons learned, successes and pitfalls in the project. Positive transformations were not limited to the business only but also included new opportunities and personal transformation for employees and a positive change of culture at the company.

512016 Tony Starck Needs Jarvis to Run Iron Man. What's the Future in AI for Continuous Process Industries?

Abstract:

Why bigdata is such a game changer? We have run plants with automation and scada levels for 30 years, and we encounter a lot of data. We visualize, we set, we adjust, we regulate. With more and more advanced technologies. We make a lot of decisions, impacting the performance of our plants. The bigdata era creates another level of opportunity. By crossing all kinds of information, we open the gate of a new worldâ€¦ of complexity. We understand that there is a gold mine awaiting our discovery, but our biological brain, the main technology we use to dig into it, is now outdated. Our brains will not crunch the bigdata! We need new technology : artificial intelligence.

Today, the most advanced continuous process plants are about to become autonomous. Transforming the decisions of process optimization to dedicated AI. This is coming fast for all of us. What does already exist? How do we implement it for what profit? What can we expect for the future? How can this benefit us?

513716 Justifying Energy Savings in Ethylene Crackers in a Low Energy Price Environment

Abstract:

Global energy prices have dropped to their lowest level in a decade and have taken with them the days of easy justification for energy improvements. In specific countries where energy prices are still subsidised, this is a familiar situation in everyday energy management. The subject of this paper is the different ways that have been used to improve the economics of energy projects in low price situations in order to make viable project investments.

Case studies will be used to illustrate such methods as:

- Debottlenecking energy critical equipment to increase throughput,
- Overcoming an energy cap to improve severity and product yield,
- Making energy savings to meet legislative targets, and
- Using energy reduction to justify changing catalysts to improve yield.

511653 Retrospective Health Impact Assessment for Pm 2.5 Pollution in Mexico City 1990-2015

Abstract:

Fine particulate matter has a well-documented association to mortality. We have conducted a retrospective health impact assessment to quantify the lives saved in Mexico City from 1990-2015 due to PM 2.5 reduction resulting from pollution control policies, utilizing the Integrated Exposure Response model developed for the Global Burden of Disease study. In addition, we monetized the health benefits using city-specific value of statistical life (VSL). Our analysis suggests that 21,000 lives saved during that period and benefits in the order of 18 billion 2015 USD. We also developed a novel model to retrospectively decompose the health benefits and attribute them to one policy utilizing sulfur reduction in fuel policy as an example. Our model indicates that this policy has resulted in 1,600 lives saved. This suggests that Mexico City air pollution policies could serve as a model for developing countries and the industry with environmental regulations at infancy stage.

512814 Sustainability Analytics - the Critical Link between Sustainability and Business Strategy for Actionable Insights

Abstract:

Advances in Enterprise Systems Are Making It Feasible for Corporations to Track and Transform Sustainability Performance. the Materiality of These Seemingly Non-Economic Impacts Is the Critical Link between Sustainability and Business Strategy. Business Analytics Enable Organizations to Convert Raw Data into Actionable Insights to Achieve Their Sustainability Goals. Sustainability Analytics Analyze the Data and Provide Insights That Help Transform Sustainability Information into Action across the Value Chain and Life Cycle. This paper will Provide Business Leaders with Insight into How to Determine Which Sustainability Metrics Are Material to Them and Relevant to Their Business, and How to Incorporate Them into Corporate Strategy, Resource Allocation Prioritization and Decision Making Process.

Based on My Book Introduction to Sustainability Analytics Under Publication By CRC Press

512839 The Reaction Mechanism and Kinetics of Magnesium Oxide for CO_2 Immobilization

Abstract:

The CO_2 immobilization is significant for the sustainable energy and chemistry industries. And the magnesium oxide is a clean and economic agent using for the CO_2 immobilization. Therefore, the reaction of CO_2 with magnesium oxide suspended in water was investigated, the mechanism proposed, and the kinetics regressed as well. The CO_2 pressures from 1 to 5 (MPa), as well as ratios of solid to liquid, stirring speeds and temperatures, significantly influenced the dissolution of magnesium oxide. And as high as 26.63 g/L MgO was dissolved in the solution under the condition of CO_2 pressure 5 MPa at 50 °C.

In the process, the magnesium oxide became magnesium hydroxide through hydration reaction, and then the hydroxide dissolved by carbonated water. In the carried out reaction of CO_2 with magnesium hydroxide suspension, no solid product was found covering the solid $Mg(OH)_2$. Therefore, a reaction mechanism based on the hydration of Mg^{2+} was proposed, where the concept of â€˜progressive liquid filmâ€™ was introduced to the reaction system. And the derived kinetic model for the dissolution was fitted the experiment data well. The pressurized magnesium hydroxide dissolution was found to be rate-controlled by the diffusion in the progressive liquid film around the solid material. The apparent activation energy, obtained from 17.1 to 25.4 (kJ/mol) in the experiments, indicated also that the dissolution inclined to be diffusion rate controlled. Moreover, the apparent diffusion coefficients of H^+ in the liquid film were calculated and an empirical expression was obtained for the first time. The kinetic model might be applicable to other dissolution reactions similarly with just soluble product.

516826 A Digital Vision for U.S. Manufacturing

Abstract:

A vision for digital manufacturing will be shared for the purpose of bringing clarity to the technology investment decisions facing your organization. The session will also cover work being done at the nationâ€™s Digital Manufacturing and Design Innovation Institute (DMDII) to advance Industry 4.0 and stand up a continuous process manufacturing test bed.

516827 Strategic Analysis of Smart Manufacturing Applications

Abstract:

A strategic analysis framework for evaluating the cost and energy impacts of smart manufacturing technologies will be introduced. The session will also cover a case study that couples intelligent digital control with process intensification (Taylor Vortex Reactor technology) to achieve flexible manufacturing of Li-ion battery cathode materials.

520695 Realizing the Digital Enterprise

Abstract:

Industry 4.0 represents the fourth wave of the Industrial Revolution. The previous industrial transformations produced dramatic changes and this one will be even more disruptive. This session will look at how productivity was actually achieved through the previous â€œrevolutionsâ€ driven by electrification and automation, then propose the lessons learned on how we can accelerate the benefits of digitalization and avoid the pitfalls which may lead to failure. Recommendations will be offered on realizing the benefits of becoming a digital enterprise based on best practice examples.

513948 Moving Towards the Smart Factory Leveraging Big Data Advancements

Abstract:

Moving Towards the Smart Factory

Leveraging Big Data Advancements

James Moyne, Michael Armacost, Mingwei Li and Amos Dor

Applied Materials—Applied Global Services

Santa Clara, CA, USA

Key Words—Big Data, Smart Manufacturing, Industry 4.0, Analytics, Prediction Methods, Predictive Maintenance, Virtual Metrology, Digital Twin, Cyber-Physical Systems Abstract Introduction

Smart manufacturing (SM) is a term "generally applied to a movement in manufacturing practices towards integration up and down the supply chain, integration of physical and cyber capabilities, and taking advantage of advanced information for increased flexibility and adaptability" [1,2,3]. It is often equated with "Industry 4.0" (I4.0), a term that originated from a project in the German government that promotes a 4th generation of manufacturing that uses concepts such as cyber-physical systems, virtual copies of real equipment and processes, and decentralized decision making to create a smart factory [4,5].

As "big data" requirements and capabilities in data volumes, velocity (rates), veracity (quality), variety (merging) and value (analytics) increase rapidly, manufacturers across industries are beginning to see opportunities to leverage big data advancements and enable the many facets of SM to improve quality, capability and throughput, and reduce cost [6,7]. The facets of SM that are aggressively being pursued include (1) leveraging advanced analytics especially for predictive technologies; (2) tightening integration up and down the supply chain network; (3) improving the use of cyber-physical system (a term used to refer to the "tight conjoining of and coordination between computational and physical resources" [8]) to improve qualities such as adaptability, reconfigurability, functionality, and reliability; (4) improving the use of real-time simulation-like tools to begin to realize the "digital twin", which refers to a digital replica of the physical system at multiple levels creating the "virtual factory" [9]; and (5) relying more heavily on a knowledge network to incorporate subject matter expertise (SME) in analytical solutions [10]. Determining the State-of-the-Art of Smart Manufacturing

While advancements are being made across these facets of SM, the progress varies from industry-to-industry and from facet-to-facet, based in large part on the pace of evolving needs in the industry and the progress of research and development in a particular facet. With respect to industry progress, the microelectronics manufacturing industry (which includes semiconductor and display manufacturing) has seen significant progress on the SM front because it is needed to address device production requirements coming from the constant innovation in areas such as mobile devices, Internet of Things (IoT) and even big data, as well as the pressures to produce high volumes, flexibly at low cost. During microelectronics manufacturing precision processes are used to create "chips" or "panels with multiple layers of complex circuitry. Each layer contains features that must be controlled to the nanometer or even molecular level. A typical microelectronics manufacturing process has hundreds of these precision control steps.

Figure 1 illustrates the SM vision for the microelectronics industry, incorporating the previously mentioned SM facets [11]. One SM area where the microelectronics manufacturing industry has excelled is in the application of analytics. As an example the dynamic quality control of these steps in this industry is facilitated with a set of capabilities collectively termed "Advanced Process Control" or APC, which includes dynamic model-based process control, Fault Detection and Classification or "FDC", and more recently predictive technologies including Predictive Maintenance (PdM), Virtual Metrology (VM) [1]. As an example of the use of cyber-physical systems, high-mix product production requiring flexibility and high throughput is addressed with real-time scheduling and dispatch capabilities. These capabilities incorporate characteristics such as product priority, production process capability, transport times and queue length, and maintenance schedules in scheduling and dispatch decisions. These and other SM capabilities make semiconductor manufacturing an idea arena with which to benchmark the current state of the art in SM.

Figure 1: A Smart Factory vision for the microelectronics industry.

In determining the progress in SM facets across the manufacturing base, it is important to understand that some SM capabilities can be implemented today, while others are still in the research phase. The latter might best be implemented today with existing leading-edge capabilities that are effective forerunners to the ultimate SM vision, rather than focusing on the ultimate solution. As an example of SM capabilities that can be implemented today, PdM is a technology that has been shown to be effective in semiconductor manufacturing. While the challenges of wide-spread PdM deployment are significant, interim solutions for PdM such as extrapolation on diagnostics data can often be effective as a complement to more complete and complex solutions, as long as prediction horizon and prediction confidence are still conveyed with the maintenance prediction event. As an example of capabilities that are effective precursors to the ultimate SM vision, the concept of the digital twin across the enterprise is a vision of factory integration in semiconductor manufacturing [6], however it is generally accepted that achieving the final vision in the industry won't happen for at least five years. In the meantime the aforementioned rules-based real-time scheduling and dispatch is an effective precursor to the digital twin. It utilizes configurable rules and workflows to capture processing and product behavior, thereby providing real-time reconfigurability decisions in order to optimize throughput in the face of quality, production schedule and product priority constraints. Further the capability is structured such at the rule base can be supported by or replaced with real-time simulation/emulation capabilities as these capabilities evolve [12].

Summary

Effective adoption of SM technology requires an understanding of the current state-of-the-art of the tenants of SM as well as the path towards the ultimate SM vision. In this tutorial the SM vision will be described including a detailed description of the individual SM tenants. The current across-industry state-of-the art of each tenant will then be detailed, identifying current solution approaches that can provide a path to the ultimate vision. State-of-the-art solution approaches and case studies in microelectronics manufacturing will be used to illustrate the SM evolution approach.

References:

[1] J. Moyne and J. Iskandar, "Big Data Analytics for Smart Manufacturing: Case Studies in Semiconductor Manufacturing," *Processes Journal*, Vol. 5, No. 3, July 2017. Available on-line: http://www.mdpi.com/2227-9717/5/3/39/htm.

[2] Wikipedia: Smart Manufacturing. Available online: https://en.wikipedia.org/wiki/Smart_manufacturing.

[3] Davis, J., Edgar, T, Porter, J., Bernaden, J., and Sarli, M. "Smart manufacturing, manufacturing intelligence and demand-dynamic performance," *Computers & Chemical Engineering*, 2012, vol. 47, pp. 145–156.

[4] Project of the Future: Industry 4.0, Germany Ministry of Education and Research, Available online: http://www.bmbf.de/en/19955.php.

[5] Kagermann, H.; Wahlster, "W. INDUSTRIE 4.0 Smart Manufacturing for the Future", *Germany Trade and Invest*, 2016.

[6] *International Roadmap for Devices and Systems (IRDS): Factory Integration White Paper, 2016 edition*. Available online: http://irds.ieee.org/images/files/pdf/2016_FI.pdf.

[7] Moyne, J., Samantaray, J., and Armacost, M., "Big Data Capabilities Applied to Semiconductor Manufacturing Advanced Process Control", *IEEE Transactions on Semiconductor Manufacturing*, 2016, Vol. 29, No. 4, pp. 283-291.

[8] *Cyber-Physical Systems (CPS) Program Solicitation NSF 10-515*, Available online: https://www.nsf.gov/pubs/2010/nsf10515/nsf10515.htm.

[9] Wikipedia: Digital Twin. Available online: https://en.wikipedia.org/wiki/Digital_twin.

[10]Armacost, M. and Moyne, J. "Moving towards the 'smart factory' in microelectronics manufacturing", *Nanochip*, vol. 12, N. 2, 2017.

[11]Hasserjian, K. "Emerging Trends in IC Manufacturing Analytics and Decision Making", Keynote, *Advanced Process Control Conference XXVIII, October 2016*. Available online via: http://apcconference.com.

[12]Lopez, F., Moyne, J., Barton, K., and Tilbury, D., "Process-capability-aware scheduling/dispatching in wafer fabs", *Advanced Process Control Conference XXIX, October 2017*. Available online via: http://apcconference.com.

513921 Future of Latent Variable Methods for Big Data Analytics in the Process Industry 4.0

Abstract:

In modern industries adopting the Industry 4.0 paradigm, the result of Industrial Internet of Things (IIoT) connecting intelligent physical entities to each other allows complex equipment units to have embedded sensors and special modules (agents) providing connection to the monitoring center. This is leading to the so-called Big Data problem in Industry 4.0.

Big data exhibit high volume and correlation, rank deficiency, low signal-to-noise ratio, changing structure, and missing values. Data in process/manufacturing industries are especially complex in structure and the useful information is often subtle.

Apart from the infrastructure (hardware) needed to manage these Big Data streams, the key point is how to analyze them to effectively extract information to give organizations new insights about their products, customers and services and steer the decision-making process. This can be particularly valuable when it is critical to maintain quality and uptime, such as in process monitoring applications, by quickly detecting and diagnosing abnormal activities, predicting the time-to-failure of equipment units or when rapid new products development is critical for company survival.

In this talk we illustrate the potential of latent variable-based multivariate statistical methods for Big Data Analytics to analyze and visualize extracted information in a way that is easily interpreted and that is useful for process understanding, real time process monitoring, fault detection & identification, process improvement, predictive maintenance, and optimization.

A discussion on the pros/cons of latent variable models with classical statistical and machine learning methods in Big Data environments will also be addressed.

512754 Advances in Utilizing Data to Improve Operations

Abstract:

Big data and Industrial Internet of Things (IIoT) are relevant topics of discussion in many commercial and industrial applications, and it also applies to the operation of refineries and petrochemical plants using analyzed data to improve operations. Online data from sensors, whether wired or wireless, can be analyzed and used to provide plant operators with the actionable information to make timely decisions for improving asset reliability and efficient operations.

Properly trained operations are invaluable in this process, as they not only act on information from analyzed data, but can also make decisions on their own based on sensor data.

Online data from sensors has been available for decades, but a transformation is now taking place due to the low cost and quick installation time for WirelessHART® (an international industry standard IEC 62591 since 2010) sensors when compared to their wired counterparts. The term pervasive sensing describes the application of wireless sensors to areas which were formerly too expensive to monitor with wired instruments, leading to the creation of big data.

Abnormal operation can quickly be identified and corrected with properly trained staff to act on this new timely analyzed information. Big data and IIoT will continue to have a role with shortages of skilled resources, thus having the staff be more efficient and effective at their role with additional information and insight into operations.

This presentation will focus on the applications where big data and IIoT have improved operations. Although the applications presented will come from oil refineries, the solutions can be applied to other continuous processing facilities.

Adaptive anomaly detection and its application to online asset monitoring

Dr Tim Butters, CMath MIMA

Head of R&D – Sabisu, United Kingdom
tim.butters@sabisu.co

Abstract

The automatic detection of faults in assets is a challenging problem, with complexities arising due to the constant changes in conditions seen in industrial operations. Here we present a method of time series anomaly identification that successfully identifies this incipient behaviour whilst adapting to the changing plant conditions. The system was applied to a hyper-compressor and used to detect packing ring failures with a high level of accuracy.

1 Introduction

Industrial plants are comprised of a large number of assets each with associated process data, much of which is digitally historised at a high frequency. This process data can be monitored either manually by engineers or automatically by software systems to identify incipient issues that need addressing. Software systems are preferable as they can monitor continuously, however, with plants constantly changing operating modes and the influences of external factors it is extremely difficult to develop a fault detection system that operators well in the general case.

Instead, algorithms that tackle specific asset failure modes are developed. Although the effectiveness of such systems can be high, the lack of horizontal applicability severely limits their use. In this work we present a method for adaptive anomaly detection in time series data that can be effectively used for asset fault detection whilst accounting for changing plant conditions.

Asset failure modes are often complex. Even with multiple examples of the same failure type there can be significant differences between the process data for each case. This, coupled with the limited number of examples of each failure mode available in an industrial setting can make data-driven modelling of asset failure using typical machine learning techniques very difficult. To overcome these issues, we have developed a method of data-driven anomaly detection that can effectively identify anomalous time series behaviour based on a small portion of past "normal" operation. This removes the reliance on large datasets associated with typical supervised or unsupervised machine learning solutions.

This anomaly detection system was used as the basis for an online monitoring solution for a two stage hyper-compressor. The specific problem being addressed is the identification and tracking of packing ring failures for the compressor to eliminate the corresponding unplanned down-time.

2 Anomaly Detection

There is no formal definition of an anomaly in the general case of time series data. For practical purposes we define an anomaly as a short period of data that does not match the behaviour of its recent past. For most asset faults the corresponding features seen in the process data fit this description, and are detectable through methods specifically designed with this definition in mind.

2.1 Algorithms

In this system several anomaly detection algorithms are used to provide flexibility in the types of abnormal behaviour that can be found. This allows the system to be uniformly deployed over different assets with minimal configuration. Below is a list of the algorithms we have implemented and tested. All of these algorithms build a statistical model of the time series, triggering anomalies when the data fails to fit this model. The data model is continually updated in real-time, which assures that when the normal behaviour of the asset changes the algorithms will adapt to this new state automatically.

AdWin

AdWin is a change-point detection algorithm developed by Bifet and Gavaldà [1]. It searches for points in the data that separate periods of behaviour based on the mean value within an expanding window. It is therefore very effective at locating changes in average value, but it is insensitive to changes in the variance of the signal.

StatWin

To capture changes in variance we developed an anomaly detection algorithm called StatWin. StatWin detects when the most recent data point lies outside of a statistical envelope defined by previous behaviour within a moving time window.

KMeans

A k-means clustering algorithm is used to model the data, and then the most recent data value is compared to this clustering to determine if it is sufficiently "close" to the cluster centres formed within a fixed window. If it is not, the value is considered anomalous.

This algorithm is based on an approach described by Hill and Minsker [2]. It effectively detects changes in both average value and variance, but rather than relying on a continuous window of allowable data it uses the cluster centroids to define regions of acceptable values. This means that this algorithm may recognise anomalies that lie within the range of previously seen data. For example, given a signal comprising entirely of the numbers 0 and 100, this algorithm would indicate a value of 50 as anomalous, as it is not sufficiently "close" to the previous values.

2.1.1 BCPD

On-line Bayesian change-point detection (BCPD) builds a statistical model of the data and marks the points at which the underlying data distribution changes significantly [3, 4]. This approach typically provides a good general purpose algorithm that is directly applicable to any time series without any changes in configuration.

2.2 Data Windowing

Data windowing provides an efficient method of dealing with continuous data in real-time. All but one of the anomaly detection algorithms, AdWin, use a fixed size moving window. AdWin is designed around the concept of a dynamically expanding window that shrinks when changes are detected to exclude the old "different" data. This means that the window expands gradually as new points are added, and then shrinks instantaneously to remove the oldest data which does not fit the current behaviour.

KMeans and StatWin require a fixed size window to be populated before anomalies can be detected. This is because accurate statistical measurements (in StatWin's case) and k-means clustering (in KMeans' case) cannot be performed without a sufficient number of data points. Once "full", new data points are added to the end of the window and the oldest point is removed from the start to maintain a constant window size.

To ensure numerical stability the on-line Bayesian change point detection algorithm also uses a fixed size window, although it does not have the limitation of requiring this window to be fully populated before anomalies can be detected. This change detection algorithm calculates the probability that each data point being analysed is a contiguous block of similar behaviour. Smaller and smaller weights are given to points more distant from the latest value.

2.3 Quorum Voting

In certain circumstances it may be appropriate to use a single anomaly detection algorithm. However, it has been found that combining the outputs of several algorithms minimises the number of false positives detected by the system. To combine the results of the algorithms a quorum voting system is employed. This means that an anomaly is only reported if the majority of the algorithms agree on the result. To account for the difference in techniques between the algorithms the results do not have to align exactly in time, but within an automatically determined timespan.

The quorum timespan can range from a few minutes to hours depending on the frequency of the underlying sensor signals and the behaviour of the asset. This accounts for the

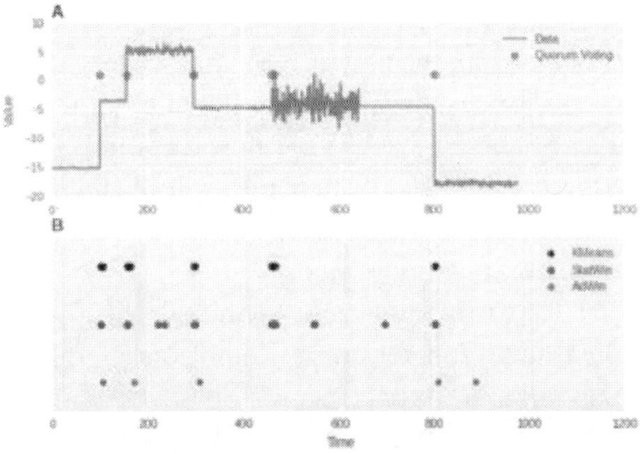

Figure 1: Anomaly detection on a test signal using the KMeans, AdWin and StatWin algorithms. The results from the individual algorithms are shown on panel B, with the signal and quorum voting result shown in panel A. From this image it is clear that the quorum voting only flags an anomaly if two or more of the three algorithms agree.

different ways in which the above algorithms detect anomalies, ensuring that the sensitivity is maintained but the false positive rate is minimised.

An example of anomaly detection is shown in figure 1. The three algorithms KMeans, AdWin, and StatWin are combined to produce the quorum voting results shown as red dots in panel A. The detection results of the individual algorithms are shown in panel B.

3 Case Study

A two-stage hyper compressor was identified as a key asset in a customers process. This hyper-compressor has 20 cylinders, each cylinder has a packing box containing 5 packing rings that are responsible for the sealing of the system. The packing boxes must be replaced periodically as over time the rings degrade and fail. If the packing boxes are not changed in time and all the rings fail, an automatic plant shutdown is triggered. This unplanned down-time lasts for approximately 2 days and costs in the region of £500,000 each time.

The packing boxes themselves are sealed and cannot be inspected, but there is a limited amount of process data available for the asset: total leak gas for the whole compressor, plunger temperature for each cylinder, and x-runout and y-runout (measures of displacement) for each cylinder. The challenge was to create a system that could monitor this process data to determine when a ring had failed, and in which of the 20 packing boxes the failure had occurred. With this information it would be possible to track the status of each packing box and schedule replacements accordingly, avoiding both premature replacement and unplanned outages due to total packing box failures.

To achieve this the anomaly detection system presented above was applied using a two-stage approach. It was found that the total leak gas measurement showed a clear spiking signature when a packing ring failed. This signature was accurately identified by the anomaly detection system with a very low false positive rate.

To identify the specific packing box containing the failed ring the anomaly detection system was then run over the cylinder-specific measurements for each cylinder (60 time series in total). Correlation analysis was then used to find any anomalous behaviour in these signals that corresponds to the anomaly found in the total leak gas. This successfully identified the packing box containing the failed ring.

The overview screen of the system is shown in figures 2, it shows a schematic of the compressor with a colour-coded health indication for each cylinder. The health is determined by the number of anomalies attributed to the cylinder's packing box, which correlates to the number of rings remaining. It is possible to have more anomalies detected than rings in the box as a ring can fail in several stages over a period of a few hours.

An example ring failure is shown in figure 3. The point at which the anomaly was detected is marked with a small Sabisu logo in the centre of the plot. The leak-gas, shown in blue on the chart, gives a clear indication of a failure. The corresponding activity in the cylinder specific measurements (yellow, green and red lines) indicates that this is the cylinder that has experienced the failure. Axis labels have been removed inline with a confidentiality agreement.

This compressor monitoring approach was found to be a highly effective method for identifying failed packing rings. As well as successfully identifying all the historical ring failures including the affected packing box, it has also found several confirmed packing ring failures in the live environment since it has been deployed to the customer site. It is now used to determine the packing box replacement schedule. Since its deployment in October 2017 it is estimated to have saved the customer £800,000 in lost production.

4 Discussion

4.1 Anomaly Detection

Here we have presented a novel method of time series anomaly detection. This method comprises of several time series analysis algorithms that build a statistical picture of the data and highlight points that do not match the picture built using the recent data history.

This method is an effective strategy for detecting anomalies whilst adapting to changing plant conditions because the data model is being constantly updated. This allows the system to implicitly learn the new "normal" state without any user intervention.

The presented algorithms are neither processor nor memory intensive, meaning that large datasets can be processed quickly and efficiently. As the algorithms build an ever changing view of the data there is no need to store entire datasets in memory for processing, allowing them to be used in an online real-time setting, and removes any limits on the size of the data that could be processed using this technique.

There is a small amount of configuration for the StatWin and KMeans algorithms, setting the threshold at which an anomaly will be flagged. These parameters are easily configured using periods of known anomalous behaviour and requires no special training. Default values for these parameters are provided

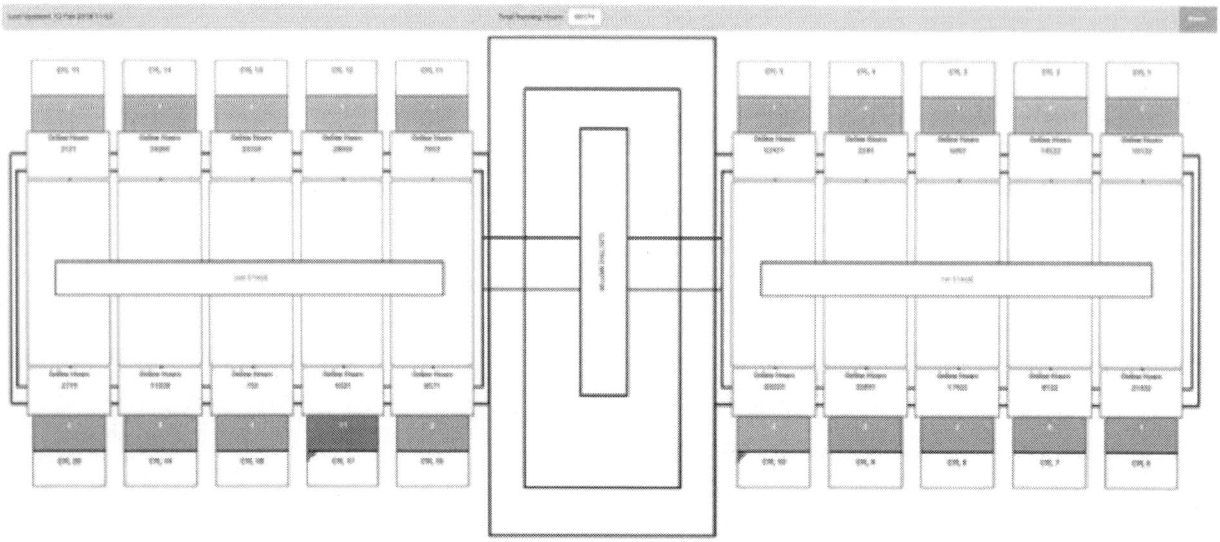

Figure 2: The overview page of the compressor monitoring system gives a clear view of asset health. Each cylinder has a colour-coded status, with the number of anomalies detected displayed for extra detail.

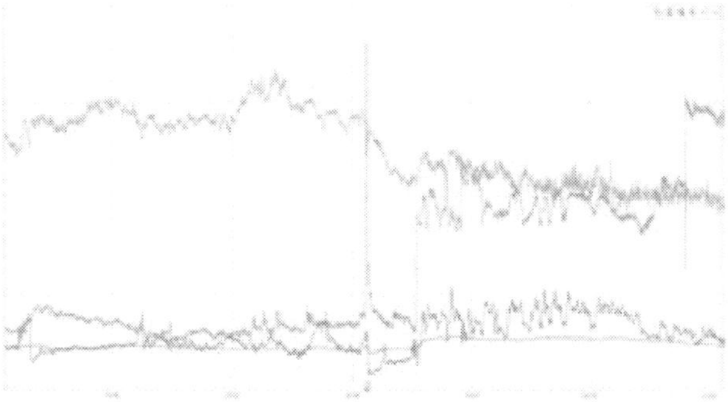

Figure 3: Data view from the Sabisu platform. This shows the compressor sensor data with a detected anomaly marked with the small Sabisu logo. The leak-gas measure shown in blue gives a clear indication of a fault, with corresponding activity in the other signals identifying this cylinder as the one experiencing the ring failure.

that have been suitable estimations in the majority of deployments of this software. It is also possible to alter the size of the sliding window used to define the data model, however, by default it is automatically determined based on the frequency of the data which has been a suitable method in all of our use cases to-date.

4.2 Industrial Case Study

As can be seen from figures 2 – 3, the presented system successfully identifies asset failures and can be used to create asset monitoring solutions. The ring failures are detected before the plant engineers had identified the fault. As well as the large number of time series to be monitored this is also partly due to the nature of this particular type of failure. As the signals quickly recover to normal conditions, if only a short section of data is assessed (*e.g.* one day) it is easy to miss a failure that happened just a few hours earlier.

The number of anomalies detected by any fault identification system is an important aspect in its "trustworthiness". As this system is designed to detect anomalous behaviour across different asset types with minimal configuration it is inevitable that there will be some false positives or missed anomalies. The preferred state is a small number of false positives with no missed true faults rather than a more conservative state in which some faults are missed. However, if the system needs to be configured such that false positives are eliminated at the potential cost of missed anomalies this is easily configured through the detection threshold parameters.

The false positive rate depends on the variability of the data being analysed. The rate in the example of compressor monitoring is extremely low, with one presented by the system over a period of approximately 6 months. In a different application of heat exchanger tube leak detection the system had a higher false positive rate, but it was still considerably lower than the acceptable rate specified by the customer in question.

5 Conclusions

An effective anomaly detection system has been presented, capable of detecting various types of anomalous behaviour across a large range of industrial assets. The system requires very little configuration, and quickly adapts to changing behaviour to allow its continued use for long periods of time without user intervention. The utilisation of quorum voting minimises the false positive rate of the system whilst retaining the sensitivity of the individual algorithms.

It has been demonstrated that the system successfully identifies packing ring failures in a two stage hyper-compressor, providing crucial information to engineers that would otherwise not be available. The system correctly identifies all historical ring failures, and since being online has saved approximately £800,000 in lost production since October 2017.

References

[1] A. Bifet, R. Gavaldà. "Learning from time-changing data with adaptive windowing", *In SIAM International Conference on Data Mining*, (2007).

[2] D. J. Hill, B. S. Minsker. "Anomaly detection in streaming environmental sensor data: A data driven approach", *Environmental Modelling & Software*, **25(9)**, pp. 1014–1022, (2010).

[3] R. Adams, D. MacKay. "Bayesian online changepoint detection", *University of Cambridge Technical Report*, (2007).

[4] P. Fearnhead, Z. Liu. "On-line inference for multiple changepoint problems", *Journal of the Royal Statistical Society: Series B (Statistical Methodology)*, **69**, pp. 589–605, (2007).

Dynamic Data Analysis™ of Large Scale Data to Monitor Fouling in Heat Exchanger Networks

Francesco Coletti[1,2*], Emilio Diaz-Bejarano[1], and Sandro Macchietto[1,3]

[1]*Hexxcell Ltd., Innovation Hub, Imperial College - White City Campus, 80 Wood Lane, London W12 0BZ, UK.*

[2]*College of Engineering, Design and Physical Sciences, Brunel University London, Uxbridge, UB8 3PH, UK*

[3]*Department of Chemical Engineering, Imperial College London, London SW7 2AZ, UK*

* f.coletti@hexxcell.com

Abstract

Data collected from heat exchangers networks in the field are typically used to monitor the thermal performance of the individual units. However, the information extracted from the data is limited in quantity and quality by the simplified models typically used in practice. Key decisions such as cleaning of heat exchangers rely on the calculation of derived quantities such as the fouling resistance which lumps together a number of factors contributing to fouling. This approach has been severely criticised in the past by various authors but it is still widely used in the industrial practice [1]. In this paper it is shown that a significantly larger amount of information and insights can be extracted from the same measurements by using rigorous models and a flexible framework. It is shown how this approach leads to a much deeper analysis of the status of the network which, in turn, helps with diagnosis and troubleshooting. An industrial case study is presented to illustrate the benefits.

Introduction

With the reduction in cost of sensors technology and IT infrastructure, the process industry is experiencing an unprecedented explosion in the amount of information being collected from each piece of equipment in each unit operation on every site. However, these data are often not exploited, particularly in the analysis of fouling in heat exchanger networks.

This paper presents an in-depth methodology that exploits existing measurements and data to monitor fouling in heat exchanger networks. The methodology, powered by Hexxcell Studio™, consists of a detailed dynamic analysis that uses historical plant data in conjunction with advanced mathematical models [2–4] to provide deep insight into the thermo-hydraulic and fouling behaviour of heat exchangers. By using this systematic analysis, it is possible to assess the quality of the measurement used, the thermo-hydraulic performance of individual heat exchangers and the entire network, the economic impact of fouling and the efficiency of cleaning actions. The analysis also allows identifying opportunities for improvement in the network (e.g. different flow split operation, cleaning strategy etc), diagnose causes of fouling and detect changes in behaviour (e.g. inorganic breakthrough) and assess the effectiveness of mitigation strategies implemented such as antifoulants, revamps etc. Combined with blending information, it is also possible to unveil key correlations between operating parameters, crude slates and fouling behaviour. An industrial case study illustrating the above benefits is presented.

Method

The dynamic analysis of the data (Dynamic Data Analysis™) is performed by using an advanced heat exchanger model [2, 3] whereby the evolution of the characteristics of the deposit is calculated directly from the plant data as a function of measured inlet conditions of temperature and flowrate, heat duty calculated from primary measurements and (where available) pressure drop measurements (a full example of this is provided in [6]). In the absence of the pressure drop measurements (which is a very typical situation in oil refineries), the deposit thickness can be calculated by fixing the deposit thermal-conductivity, allowing the calculation of hydraulic variables under reduced flow area conditions due to the presence of the layer: fluid velocity, shear stress, pressure drops, etc.

Case Study

The case study focuses on the hot end of the preheat train of a mid-size oil refinery, comprising of a total of five heat exchangers. The data analysis was performed on a set of plant measurements (temperatures, flowrates) for an operating period of ca. 1200 days.

Results

The analysis provides a comprehensive insights into the evolution of fouling, the performance of the exchanger, operating conditions and dimensionless numbers of interest on both tube and shell sides. Here, we focus on some of the most relevant findings for this case study.

First, the evolution of the fouling behaviour was obtained for the entire period in a seamless simulation. For the sake of comparison, the overall heat transfer fouling resistance for three of the units is shown in Figure 1. The analysis helped quantify the efficiency of intermediate cleaning actions carried out at the refinery. Cleaning actions are indicated in Figure 1 with labels M for mechanical and C for chemical and are shown as sudden drops in the fouling resistance. In this case the analysis also helped identify differences in fouling behaviour between various units. The three units in Figure 1 share very similar tube-side design and operating conditions. However, the fouling resistance in E04 is approximately twice that in E02AB and E05AB indicating a possible shell-side fouling build-up in E04 which does not affect other units [7].

By considering tube-side fouling in all units, and a 50/50% distribution of fouling between tube and shell sides in E04, the analysis provided the thermo-hydraulic evolution of the performance of the various units shown in Figure 2a. The thermo-hydraulic plot (TH plot) displays the ratio of the fouled-to-clean heat duty against the ratio of the fouled-to-clean tube-side pressure drop.

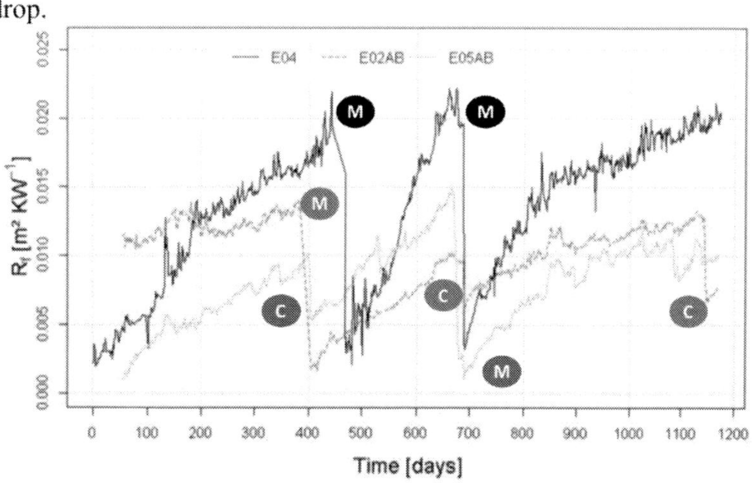

Figure 1. Total fouling resistance in E02AB, E04 and E05AB. M and C indicate mechanical and chemical cleaning, respectively (adapted from [7]).

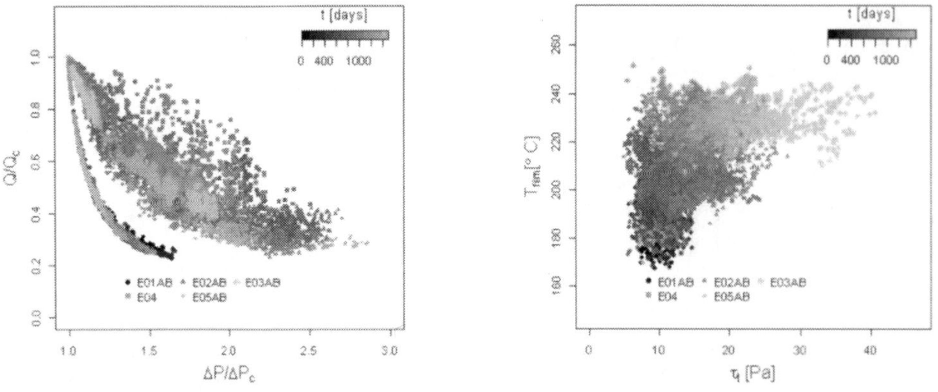

(a) (b)

Figure 2. Tube-side TH-plot (a) and tube-side film temperature against shear stress (b) under fouled conditions over a period of ca. 1200 days.

As fouling builds up, heat transfer gradually degrades and the cross-sectional flow area is restricted, which decreases the heat duty and increases the pressure drop. Simultaneously, the analysis provides the operating conditions at each point of the exchanger under fouled conditions. Figure 2b shows the tube-side film temperature against the tube-side shear stress. The figure shows that these conditions, which are typically used to characterize fouling behaviour, vary significantly during the operation. This variability is result of a combination of process variability and fouling build up. The latter can be visualized as the gradient from darker points (older in time) to paler points (latter in time). Figure 2 also shows that E04, with the fouling distribution selected, indeed operates under similar conditions and performance than E02AB and E05AB.

Finally, an extension of the method, the Fouling Propensity Analysis™, can be used to estimate fouling rates and find correlations between operating conditions and the concentration of the oils processed. This allows identifying oils that are particularly problematic with respect to fouling deposition or, conversely, oils that have a beneficial effect and help reducing the fouling resistance. Figure 3a shows the results of the correlation analysis between fouling rates and the mass fraction of the oils processed in the refinery. The information on oil composition has been modified to protect confidentiality without affecting the significance of the example presented.

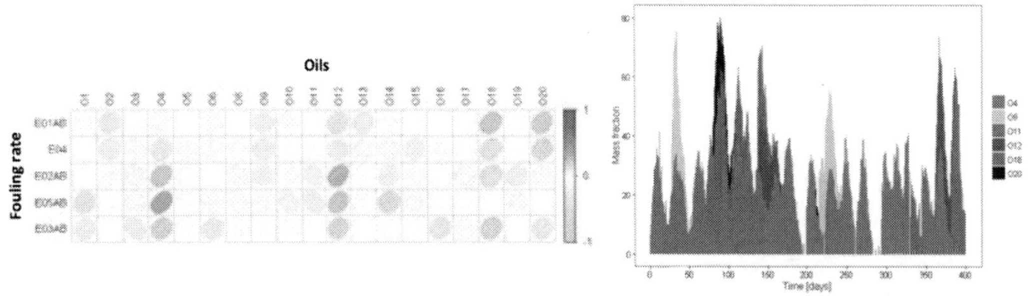

(a) (b)

Figure 3. Result of correlation analysis between fouling rates and mass fraction of the oils processed (a) and processing history of oils identified as problematic (b). Information on oil composition has been modified to protect confidentiality without affecting the significance of the example presented.

Conclusions

An industrial case study has been presented in which Dynamic Data Analysis™ allowed studying the evolution of the fouling behaviour and its impact on the operating conditions inside the tubes (film temperature, shear stress) and the overall thermal and hydraulic performance of the exchangers. The analysis allowed identifying and quantifying the extent of cleaning actions and detecting the different fouling behaviour in one of the units, likely due to severe shell-side fouling. Finally, the fouling rates obtained were cross-correlated to information available on the crude oil slates processed. This permitted identifying oils particularly problematic towards fouling deposition.

References

[1] F. Coletti and G. F. Hewitt, Crude Oil Fouling: Deposit Characterization, Measurements, and Modeling. Boston: Gulf Professional Publishing, 2014.

[2] F. Coletti and S. Macchietto, "A Dynamic, Distributed Model of Shell-and-Tube Heat Exchangers Undergoing Crude Oil Fouling," *Ind. Eng. Chem. Res.*, vol. 50, no. 8, pp. 4515–4533, Apr. 2011.

[3] E. Diaz-Bejarano, F. Coletti, and S. Macchietto, "A new dynamic model of crude oil fouling deposits and its application to the simulation of fouling-cleaning cycles," *AIChE J.*, vol. 62, 1, pp. 90–107, 2016.

[4] E. Diaz-Bejarano, F. Coletti, and S. Macchietto, "Impact of Complex Layering Structures of Organic and Inorganic Foulants on the Thermohydraulic Performance of a Single Heat Exchanger Tube: A Simulation Study," *Ind. Eng. Chem. Res.*, vol. 55, no. 40, pp. 10718–10734, 2016.

[5] E. Diaz-Bejarano, F. Coletti, and S. Macchietto, "Thermo-Hydraulic Analysis of Refinery Heat Exchangers Undergoing Fouling," *AIChE J.*, vol. 63, no. 3, pp. 984–1001, 2017.

[6] E. Diaz-Bejarano, E. Behranvand, F. Coletti, M. R. Mozdianfard, and S. Macchietto, "Organic and Inorganic Fouling in Heat Exchangers – Industrial Case Study: Analysis of Fouling State," *Appl. Energy*, vol. 206, pp. 1250–1266, 2017.

[7] E. Diaz-bejarano, F. Coletti, and S. Macchietto, "Modeling and Prediction of Shell-Side Fouling in Shell-and-Tube Heat Exchangers," *Heat Transf. Eng.*, vol. Accepted f, no. https://doi.org/10.1080/01457632.2018.1446814, 2018.

512443 Machine Learning Collective Variable Discovery for Materials Design and Engineering

Abstract:

Data-driven modeling and machine learning have opened new paradigms and opportunities in the understanding and design of soft and biological materials. The automated discovery of emergent collective variables within high-dimensional computational and experimental data sets provides a means to understand and predict materials behavior and engineer properties and function. In this tutorial, we will describe two machine learning techniques for collective variable identification â€" nonlinear manifold learning using diffusion maps, and nonlinear dimensionality reduction using autoencoding neural networks (â€œautoencodersâ€). We will provide illustrative applications of diffusion maps to recover low-dimensional assembly landscapes for self-assembling patchy colloids from simulated and experimental data, and inverse building block design by rational sculpting of the landscape to stabilize desired aggregates. We will provide illustrative applications of autoencoders to identify collective variables for protein folding and the use of these coordinates to accelerate molecular dynamics simulations.

512055 Big Data Analytics for Upstream Processes

Abstract:

Modern industries are awash with large amount of data. Extraction of information and knowledge discovery from data, particularly from day by day routine process operating data, is especially challenging. There exist numerous challenging issues such as data nonlinearity, non-Gaussian distributions, high dimensionality, collinearity, multiple modal operations, outlying points, missing measurement etc that must be considered during the information extraction process. This presentation will discuss state-of-the-art development of data analytics to deal with these issues and to develop predictive models, soft sensors, fault detection and diagnosis monitors, and optimization from data for upstream processes, particularly oil sands processes. The concept of data analytics is illustrated in detail by a number of applications. The success of big data analytics in industrial applications will be elaborated.

ENTERPRISE MANUFACTURING INTELLIGENCE TO BRING DATA AND KNOWLEDGE TOGETHER FOR REAL TIME DECISION MAKING

Chaitanya Khare
Associate Analytical Manager
The Dow Chemical Company

Mary Beth Seasholtz
Associate Technology Director
The Dow Chemical Company

Richard Rolke
Global Improvement Leader
The Dow Chemical Company

Jim Petrusich
Vice President of Sales
Northwest Analytics

Prepared for Presentation at the 2018 Spring National Meeting

Orlando, FL, April 22-26, 2018

ENTERPRISE MANUFACTURING INTELLIGENCE TO BRING DATA AND KNOWLEDGE TOGETHER FOR REAL TIME DECISION MAKING

Chaitanya Khare
Associate Analytical Manager
The Dow Chemical Company
Mary Beth Seasholtz
Associate Technology Director
The Dow Chemical Company
Richard Rolke
Global Improvement Leader
The Dow Chemical Company
Jim Petrusich
Vice President of Sales
Northwest Analytics

Abstract:

Enterprise Manufacturing Intelligence (EMI) is a method which gives personnel quick access to critical parameters in real time and provides suggestions on pro-active actions to take to keep the plant running well, therefore avoiding plant crisis. Knowledge about a plant or unit operation is captured through EMI in two ways: (1) trending representative tags and calculations in real time and (2) providing suggestions for pro-active actions to take based on the deviation from normal conditions.

The future of the manufacturing industry is towards building intelligence into an organization through use of data -driven decisions. A significant part of this process is a onetime investment to capture current technical knowledge of the operation. This establishes a solid platform for ongoing continuous improvements. The value that can be gained by putting the data into context is through improved reliability of the process.

The future in the manufacturing industry will include a knowledge database which is connected to the EMI tool. When real time data are connected to Dow's immense collection of knowledge through different sources of literature, this should result in a system which pro-actively engages subject matter experts to prevent any unforeseen circumstances like a plant shutdown while at the same time evolves itself with the new learnings from each new situation that are encountered.

INTRODUCTION

The chemical industry has come a long way since its manufacturing inception in 1736 with Sulfuric acid production [1]. Since then manufacturing of chemicals has increased in capacity and efficiency in the conversion of feedstocks. Improvements are a constant in the chemical industry, with safety of the people, community and the environment remaining the number one priority.

While technology exists for process control and optimization, efficient troubleshooting of a problem/issue is still a challenge. Traditional troubleshooting projects start off with first understanding the plant status by pulling and assessing the plant data. This is followed by calculating pertinent metrics, and following them back in time to understand the series of events that caused the current problem.

Enterprise Manufacturing Intelligence (EMI) is a method which gives personnel quick access to critical parameters in real time and provides suggestions on pro-active actions to take to keep the plant running well, therefore avoiding plant crisis. Knowledge about a plant or unit operation is captured through EMI in two ways: (1) trending representative tags and calculations in real time and (2) providing suggestions for pro-active actions to take based on the deviation from normal conditions.

This report is a guiding document for how to bring data and knowledge together for real time decision making. This report will be particularly useful for the people who are charged with keeping plants running well; whether that be in a manufacturing plant, R&D, or the technology center.

WHAT IS EMI?

Enterprise manufacturing intelligence (EMI) is a term which applies to software used to bring a corporation's manufacturing-related data together for the purposes of reporting, analysis, visual summaries, and passing data between enterprise-level and plant-floor systems. As data are combined from multiple sources, it can be given a new structure or context that will help users find what they need regardless of where it came from. The primary goal is to turn large amounts of manufacturing data into real knowledge and drive business results based on that knowledge [2].

At Dow, EMI contains several pieces, (see Figure 1). It includes: (1) real-time status board about data and pertinent calculations, (2) easy drilldown to graphs, (3) suggested actions for keeping the plant in the "sweet spot", and (4) place to enter comments, ideas, collaboration which can be mined in the future. That is, it becomes an "Operational Guidance System" useful for collaboration between Plant, R&D and Technology Center personnel. The information readiness for decision making is key for helping any manufacturing plant by the technology center and R&D. The information readiness makes EMI a meeting place of all the real time decisions in a startup, normal, or shut down phase of an operating unit.

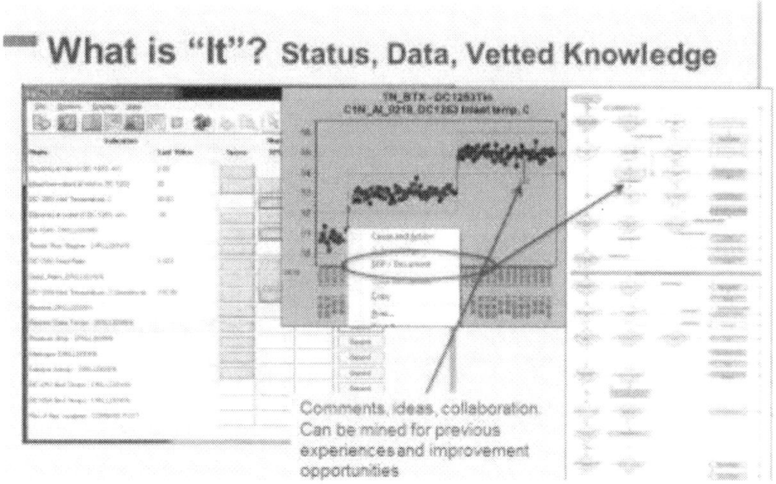

Figure 1: Components of an EMI Deployment

HOW TO IMPLEMENT EMI

To understand how to implement EMI it is good to first understand how plant support has been historically conducted.

Old paradigm:

Within the old paradigm, the plant would call out for help to either the tech center or R&D organization after they were already having problems. (see Figure 2). Then, a team would go through the rigor of first understanding the issue by speaking with the plant operations team. They would then start collecting the data for the plant. They would analyze this information while at the same time testing their different hypothesis. Eventually something would be learned and a report would be written about the knowledge gained. But, the learnings would likely not be sustainably incorporated into plant operations, so problems recurred.

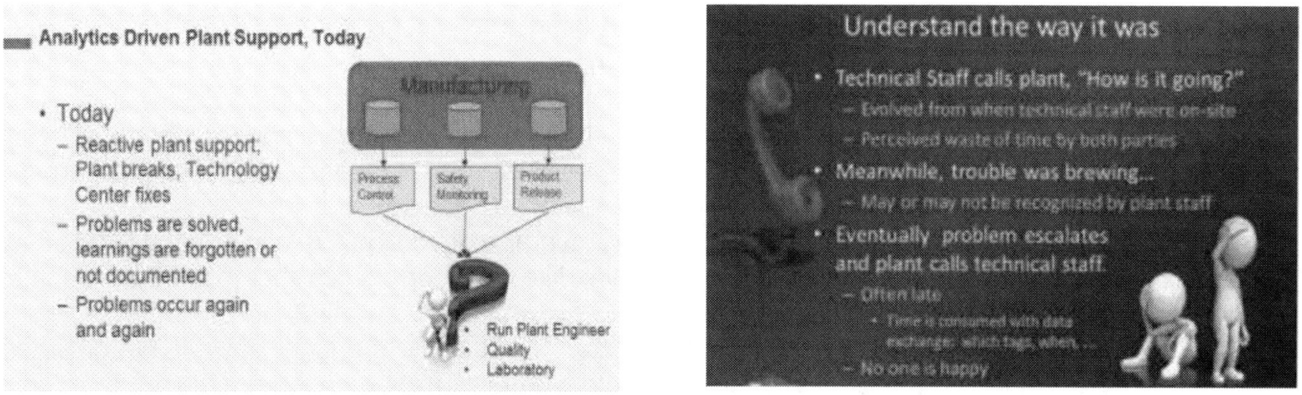

Figure 2: Old paradigm of plant support

New paradigm:

The new paradigm is to implement EMI to provide an early warning system which will prevent issues from occurring and sustainably incorporate learnings into plant operations. (see Figure 3). The target selection process for EMI implementation is based on a business's prioritized AU gaps. Choosing projects in this way leads to good support from the leadership and the technical experts.

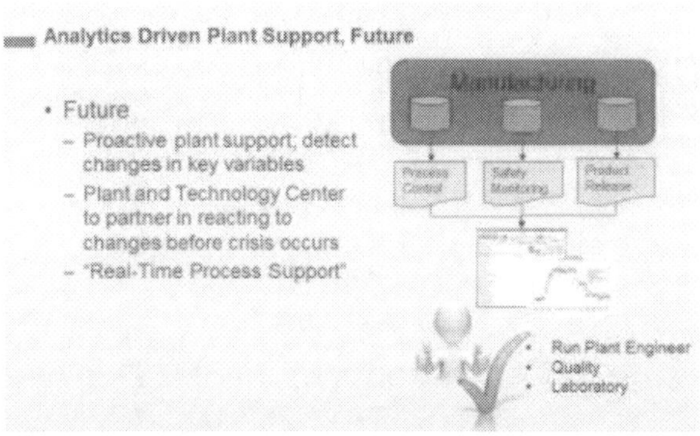

Figure 3: The new paradigm using EMI for plant support

Once a priority AU gap is identified, a team of subject matter expert's (SME's) from the technology center and/or business R&D and the operating plant is formed. This team is charged with building a "logic decision map". Rigorous brainstorming by this team is facilitated to have a healthy and positive conversation. During the initial step the goal is to draw a simplified process flow diagram and collect the most important parameters in the plant. As discussions proceed about the different parameters, historical and real time data are brought into the conversation. This helps to test different hypothesis around the parameter of choice. These data-driven discussions help expose and if needed, debunk, tribal myths.

Once the process is understood by the team, and there is an appreciation for the varied metrics highlighted, the next step is to discuss the AU-gap. For example, "What, when not monitored, would shut the plant down?" This question can trigger a lot of different ideas from the SME's. Below is an example of the interactions described:

- Assume that the overall production goal in a plant is the conversion across a series of reactors. The target is to keep it above 90%.
- Next, ask the question, "What, when not monitored, would shut the plant down?" Discussion on the above question reveals that factors other than failing conversion can bring the plant down. That is, the overall conversion is not the most critical parameter as it can be tuned to go above 90% if some activity is lost in the reactors/catalysts.
- So what would be more relevant than the overall conversion? It could be quality of the feedstock, it could be impurities, or something like water or oxygen that may be a poison for the system.
- Where do we get the answers to the question "What shuts the plant down"? We get these answers quite often from a previous root cause analysis from a previous situation that led to a plant shutdown. These answers could be known from the catalyst provider or from the technology MET documents or R&D work.

Now the team should be able to build out the logic decision map as they now understand the process and what causes the AU Gaps. That is, the team needs to identify the most important parameters to monitor, what ranges those metrics need to be kept within, as well as what actions should be taken to keep the key metrics in range. When thinking through appropriate actions, the team must resolve how to balance competing phenomena. All this information is put into a tool which shows real-time data as well as suggested actions to take if metric(s) are out of range. The data should be arranged so that at first glance only the pertinent metrics are shown, but with a few clicks, other data which are helpful for understanding a situation more completely can be found.

What makes the new paradigm better than the old?:

The new paradigm of creating an EMI system helps the team focus on analysis of data and determination of pertinent metrics without a plant crisis looming over head. In the old process, the troubleshooting would start after the trouble started, which was too late. With regards to obtaining and interacting with the data in order to proactively prevent a crisis from developing, the EMI approach allows for more time to be spent analyzing the data than in the past (see Figure 4).

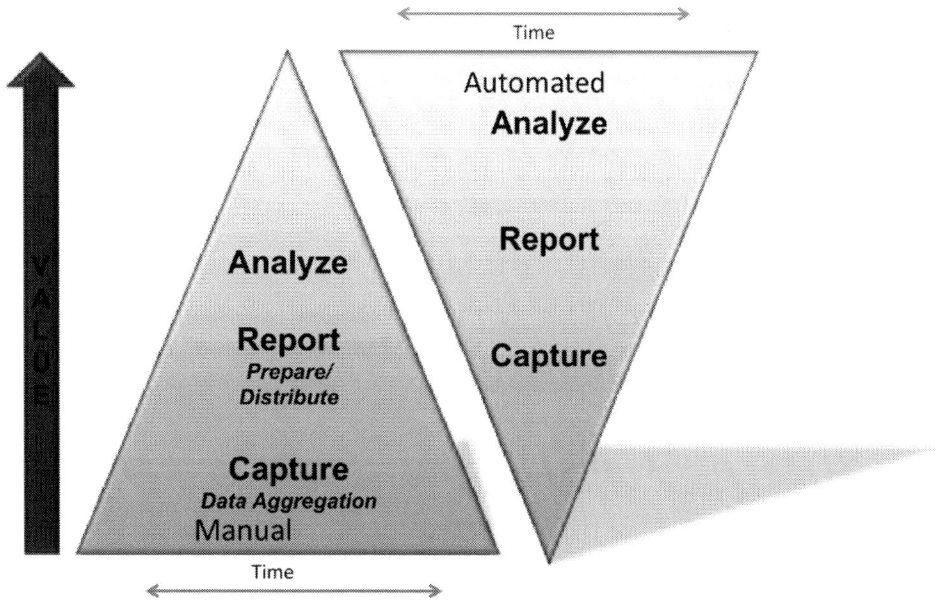

Figure 4: Difference between the old and the new paradigm in terms of time spent vs creating value

One can apply EMI for making decisions at different levels within an organization (See Figure 5). In terms of implementing EMI at the transactional level, one should be aware that at Dow a number of systems are already in place, and this is likely not an opportunity-rich environment.

The most valuable EMI implementations to date have been at the tactical level of operation. The tactical level is where plant support takes place and the different business functions work together to help the plant troubleshoot or avoid a crisis. At the tactical level decisions are made taking into consideration the details within the technology and the current operating condition of the facility. Having a logic decision map at this level of operation delivers the most promising results and facilitates collaboration between the support and operations functions. That is, EMI becomes the meeting place for R&D, technology center and the plant to collaborate to proactively prevent a problem.

The operational level EMI implementation shows the summary of the entire plant operation encompassing multiple technologies. Such an implementation is very useful during morning production meetings as it provides an efficient summary of the different plant areas. Time is spent talking about how to proactively prevent problems, not fussing over what data to look at. This helps to improve the efficiency of the plant staff meetings with data readily available for decision making. Another kind of operational level implementation is to show the status of same technology (ex. reactors) across a fleet of plants. Then decisions can be made about which locations operate more or less efficiently and learnings can be leveraged across the fleet.

The strategic level of EMI implementation is immensely valuable for making large scale capital decision across a business or an organization. The information presented can provide guidance in making decisions at the business level. Making such information available in real time is useful to the business leadership driving the decisions.

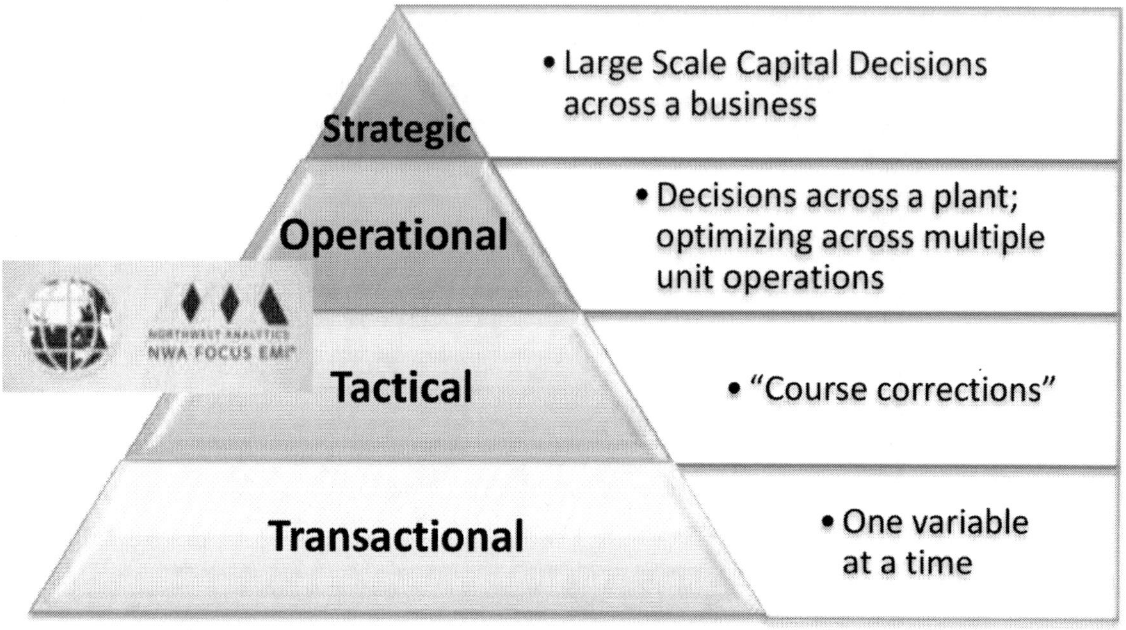

Figure 5: Decision matrix with different tools at different stages

Managing the change in culture:

An important question generally asked is: "How does a typical implementation proceed?" (See Figure 6). Building the logic decision map is a major step in implementing EMI, however it is only half the journey. The second half of the journey is more complex than debating the technical details. This part of the journey is the cultural transformation from reactive troubleshooting to proactive data-driven decision making.

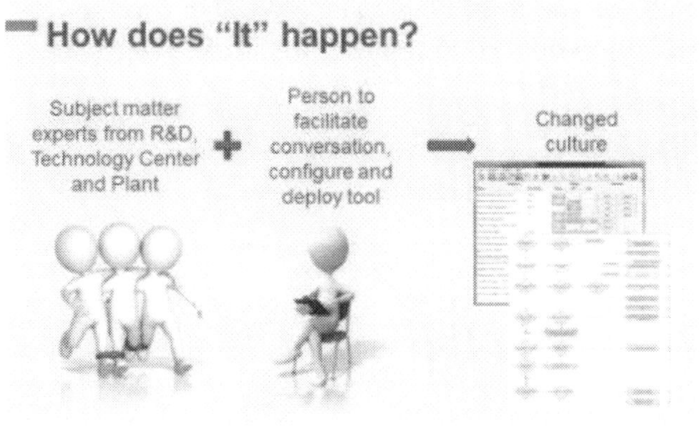

Figure 6: Typical method of EMI implementation

Users of EMI come from the Technology Center, Plant, Lab, Quality, and R&D. Implementing EMI helps each user from the various functions connected with the production facility. The production or improvement engineer in the plant using EMI gets the information about his/her plant, perhaps in the first 5 minutes of their day. This lets them head into meetings with knowledge about the plant status; therefore they can answer questions that may arise. In addition, knowing the current status of the plant helps ensure the plant prioritizes the day according to the situations that have been observed in EMI.

Users from the Technology Center and R&D are looking to understand what is ongoing at all the plants within a particular technology. With EMI in place across a business for that technology, a technology SME can have accurate information about all plants, even if he/she is located remotely from the manufacturing facilities. This information helps prioritize efforts in the technology center and R&D.

The Lab and Quality EMI implementations put in perspective all the online/offline measurements in the plant which maintains an accurate focus on helping understand the results from the process to maintain the expected product quality.

The first draft of the EMI deployment is often all red (See Figure 7 – Start). That is, many of the parameters are initially operating outside of the recommended ranges. As the users start using EMI every day, they monitor the newly established key parameters. They also learn what appropriate changes/improvements are needed to keep these key parameters in acceptable ranges and the system becomes green (See Figure 7– Now). This transformation from the start of the implementation to end of the implementation is the culture change taking place from reactive to proactive. But, it is not an easy step. This change takes time, energy, and a lot of commitment to doing the right thing from the plant, technology center and R&D. However, at this point in the implementation the data are starting to work for us, instead of us working to acquire and assemble the data. The critical parameter control charts will start informing users of any deviations from the normal and the decision map will guide users in making the appropriate decision in a proactive manner.

When EMI is implemented in plant operations, the above mentioned users benefit immensely. For example: In a morning production meeting, usually there is the main day staff which is required. The day staff in a large plant is around 30 people. Typically, to go through all the details in a big facility the morning production meetings could span around one hour. Discussions take place across several issues ongoing in the plant. However, we might be spending time from folks who are not really necessary to be present in those discussions. With the use of EMI, the morning production meetings can be reduced to a mere 15 minutes because the data are organized and readily available. EMI helps in maintaining focus on safety and the current issues that are alarming for the plant. After the larger team sees the issues, it is possible to take the issues offline to discuss in a small SME group. This approach can save a lot of time for several of the day staff in an operating facility. In simple terms, time saved would be 45 minutes for 30 people every day.

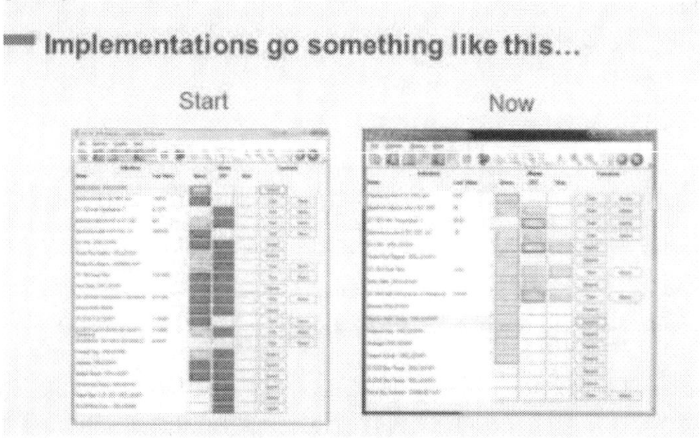

Figure 7: Natural progression of an EMI implementation from start to the current state

FUTURE OF EMI

The future of the manufacturing industry is towards building intelligence into an organization through use of data -driven decisions. A significant part of this process is a onetime investment to capture current technical knowledge of the operation. This establishes a solid platform for ongoing continuous improvements. The value that can be gained by putting the data into context is through improved reliability of the process (See Figure 8).

The future in the manufacturing industry will include a knowledge database which is connected to the EMI tool. When real time data are connected to Dow's immense collection of knowledge through different sources of literature, this should result in a system which pro-actively engages subject matter experts to prevent any unforeseen circumstances like a plant shutdown while at the same time evolves itself with the new learnings from each new situation that are encountered (see Figure 9).

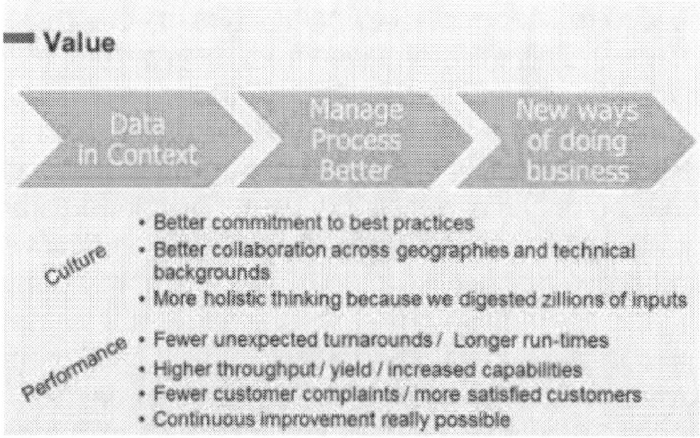

Figure 8: Return on investment from using our data through the intended culture and performance

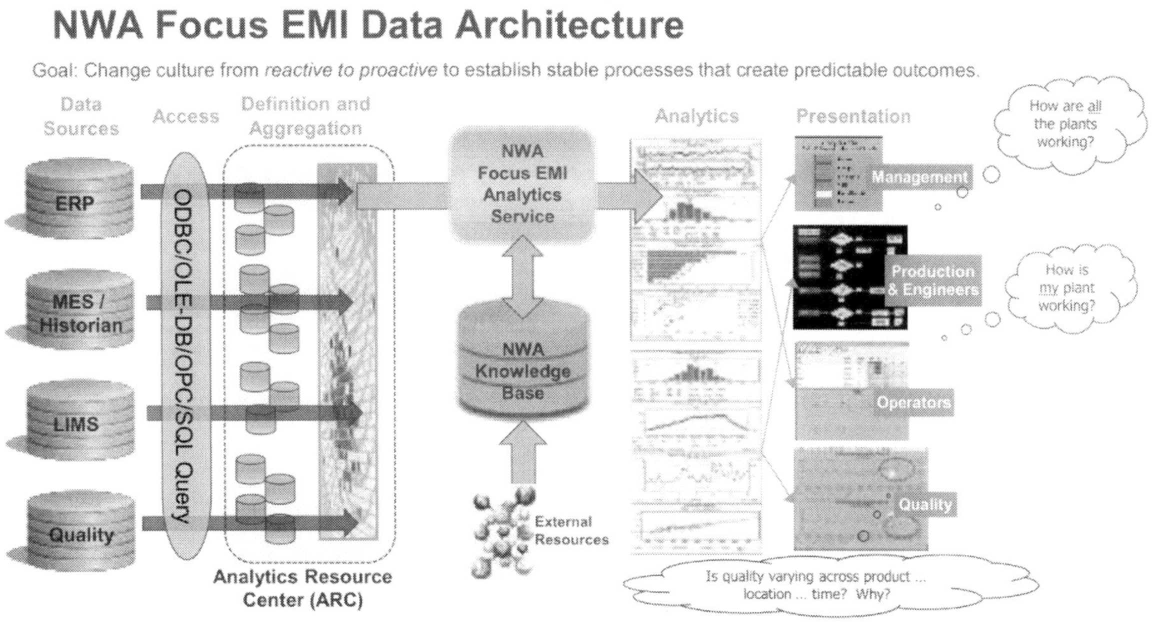

Figure 9: Architecture flow chart for generating the knowledgebase through the use of data in NWA Focus EMI

ACKNOWLEDGMENTS

We would like to acknowledge numerous discussions during the germination of EMI with Lloyd Colegrove, and Howard Clark. We would also like to acknowledge the Hydrocarbons business in helping us test this new concept as a pilot and staying on with us during the initial phase.

REFERENCES

1. First industrial revolution, https://en.wikipedia.org/wiki/Industrial_Revolution
2. EMI, https://en.wikipedia.org/wiki/Enterprise_manufacturing_intelligence
3. L. Colegrove, M. Seasholtz, C. Khare (2016), "Getting started on the Journey", Chemical Engineering Progress, 112(3), 41-45.
4. C. Khare, S. Chin (2017), "Potential for data analytics opportunities in SMART chemical industry", Chemistry Today, March/April, vol. 35(2).

AIChE Paper Number: 169B

Big Data Analytics for Improved Overall Asset Effectiveness of Ethylene plants

Dr. Pratap Nair
President & CEO
Ingenero

Prepared for Presentation at the 2018 Spring National Meeting
Orlando, Florida, April 22-26, 2018

Big Data Analytics for improved Overall Asset Effectiveness of Ethylene Plants

Starting up and sustaining production levels consistently from an ethylene plant does not always match expectations. Maintaining and managing furnace operations, the Cracked Gas compressors, Transfer Line Exchangers, chiller health and the distillation operations in the downstream cold section, to get the best Overall Asset Effectiveness (OAE), is a constant challenge. OAE is a combination of Throughput, Availability and Yield. Any one of these three parameters could impact performance of the plant. Big Data Analytics can provide insights and prescriptions that help balance these three parameters to sustainably maximize the OAE.

Data availability is not a constraint, but utilization of the available data is. Data capable of providing insights is available from diverse sources like, the DCS, operator logs, laboratory, maintenance and engineering documents. Traditionally about 90% of the collected data is archived and not even looked at. More complete utilization of this data, utilizing Big Data Analytics techniques modified to suit the process manufacturing application, yields valuable insights to help improve the overall asset effectiveness as well as individual equipment effectiveness.

While several Ethylene plants have used subsets of these big data analytical techniques quite successfully to enhance OAE, the full application of the big data techniques tailored to process manufacturing allows improved OAE levels beyond the current paradigms.

Successful implementations have shown that Big Data Analytics in Process Manufacturing is not just an off the shelf tool that can be picked up and applied. Success of the analytics initiative for Chemical Process Manufacturing entails several aspects, to ensure effective implementation and sustenance. Effective data handling, a synergistic use of Machine Learning and fundamental models for Descriptive, Predictive, Prescriptive and Diagnostic applications, an effective business intelligence interface to provide readily actionable visual information based on a solid data and analysis foundation and a work process are all important aspects to be considered.

Data quality issues and skills availability are two of the top challenges organizations typically face when trying to implement analytics systems. The paper shows how these challenges were successfully addressed using combined models based analysis and the Internet of Things (IoT).

This paper showcases ethylene facilities where a Big Data Analytics initiative has been successfully implemented at the process manufacturing level to improve OAE. The paper also covers the key ingredients and success factors based on lessons learnt from these applications.

The Need in Ethylene Plants

Starting up and sustaining production levels consistently from an ethylene plant does not always match expectations. Typical challenges faced when operating an ethylene plant:

- Furnace run lengths below expectation
- Unexpected tube failures and lower than expected tube life
- Rapid coking
- Inefficient furnace decokes
- Equipment reliability and lower than expected uptime
- Below capacity operation
- Recurring fouling in TLEs, chillers, CGC circuit
- Utility limitations
- Refrigerant and caustic losses
- Lower than expected product recovery
- Higher than expected specific energy consumption
- Off spec product
- Excessive flaring

The raison d' être for the ethylene plant asset is to get best possible production of ethylene, without excessive costs and without incurring safety and compliance incidents. Each of the challenges listed above, impact the production from the asset (the ethylene plant).

The ethylene plant is complicated to run, with all its parameter interactions. Most of the ethylene plants today, have large amounts of data and more, that is collected and available, indicating what is going on within the pipes. Making sense of this data can give insights into why the plant is behaving the way it is and hence enable actions to fix any anomalous behavior, that threatens a manufacturer's profitability and survival, be it efficiency, safety or regulatory compliance related.

So how does one use the data to address the need? While the underlying approach to address this need may sound familiar, the level to which the inaccuracies in the data are handled and gross approximations made and/or intuition relied upon to address these needs using data, would vary depending on the decade we are operating in. Digital technologies that are now maturing can be leveraged to help minimize inefficiencies and risks that may have been considered acceptable operations norms before.

Using data - The overall approach

The overall approach to define metrics using the data from the plant operation, to better understand and characterize the operation and serve as a basis for actions to saftely improve performance, may not very different from prevalent practice. The digitalized approach improves the timeliness and reliability of the information, allowing a much higher value to be derived than before. The overall approach entails:

- Defining metrics that help measure the effectivess of the asset (the plant) in meeting its production goal
- Identifying and defining key parameters that impact the effectiveness metric
- Calculating the metrics and key impact parameters
- Defining performance benchmarks
- Computing the gaps from possible performance benchmarks
- Monitoring the metrics and parameters
- Identifying gaps from benchmarks
- Zeroing in on the root causes for these gaps
- Developing actions to close the gaps
- Implementing the actions
- Tracking impact of actions on the gaps
- Fine tuning and keep up with changes caused by variables we have no control on, with the help of these metrics

The Overall Asset Effectiveness (OAE) is a useful metric that provides a high level view of the asset performance. We can drill down from this metric to details at individual equipment of device levels, to understand the root causes of below expected OAEs. The effectiveness of individual equipment or sections can also have an analogous metric: Overall Equipment Effectiveness (OEE). This metric has been widely used in discrete manufacturing and in Batch process manufacturing to some extent, but has not been used as much in contnuous process manufacturing, especially in bulk chemicals manufacturing.

Overall Asset Effectiveness (OAE) – A useful metric

The overall asset effectiveness (OAE) is a factor comparing the actual production of the product to a benchmark production. In case of the ethylene plant it is tons of ethylene produced over a specified period of time (could be an hour, a day, a week, a month, year,....), as compared to a benchmark production of the product over the same period. Product could also be defined as Ethylene + Propylene, depending on the key objective of the plant. Possible benchmarks:

- Design production
- Historical Best production
- Ideal possible production

It is more convenient to take design as the basis benchmark for calculating the OAE, since it is relatively more constant than the other two benchmarks. The other benchmarks can then be normailzied against the design basis.

There are three key parameters that impact the production:

- Throughput (total feed to the plant, e.g. kgs/hr of Ethane)
- Yield (kgs of product/kgs of feed)
- Availability (onstream time of the plant)

Each of these parameters when compared to the corresponding design benchmark, gives rise to three effectiveness factors: Throughput Effectiveness (TE), Yield Effectiveness (YE) and Availability Effectiveness (AE). The formulae and relationships are detailed in Annexure 1.

Two case examples of the OAE metric and its key impacting parameters (TE, YE and AE) are shown in Figure 1 below.

Figure 1

Overall Asset Effectiveness Case Examples

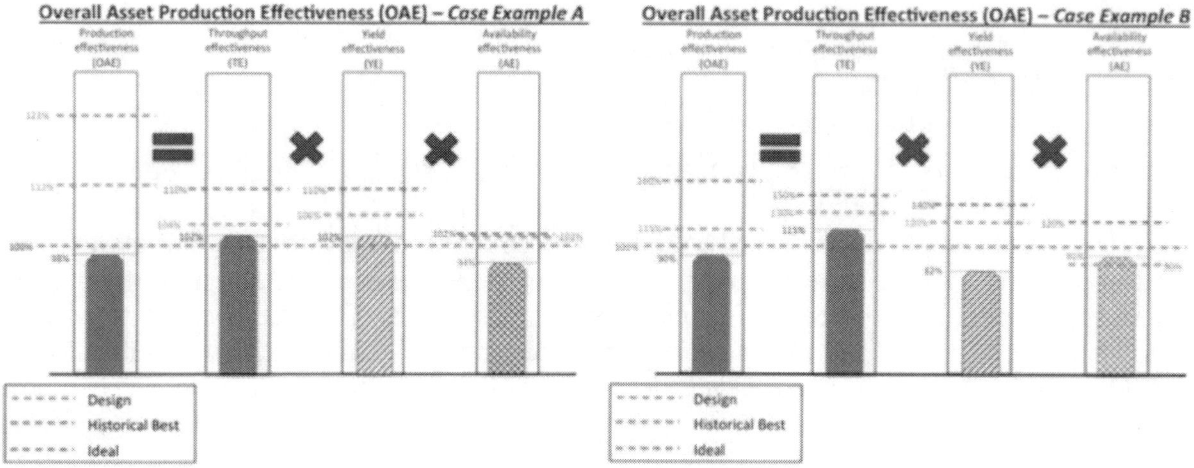

The gaps could also be shown as waterfall charts (Figure 2 below), indicating the extent to which each parameter (TE, YE and AE), impacts the OAE.

Figure 2

Overall Asset Effectiveness Gap Analysis

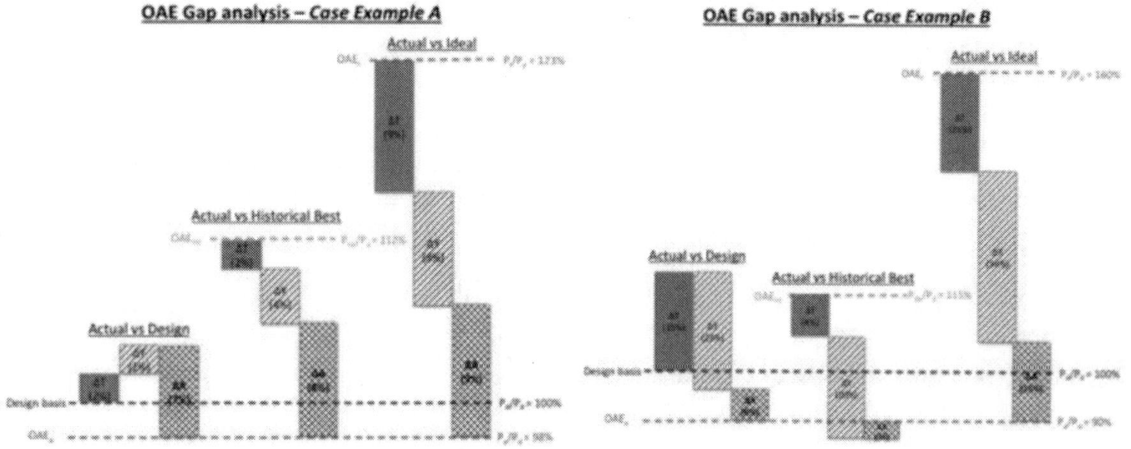

In the Case Example A shown in Figures 1 and 2, the OAE is 98%, meaning that the actual production for the period considered is at 98% of the production the plant has been designed for. The figure also shows that the plant has been operated before at 112% of design production levels (The Historical Best), for the very same conditions of the feed profile and furnace feed configuration. Also, at these feed and furnace configurations, calculations show that the plant could possibly be run at 123% of the design production levels, at the existing conditions of the plant equipment.

The resolution of OAE into its components (Throughput, Yield and Availability) shows that the main cause for production being below design levels, is on account of the plant availability during this time period considered, which is at 94% of design availability. The throughput and Yield are operating above design levels.

While the details have not been shown here in the diagram or as separate figures, further drill down in this case showed that this availability short fall came from lower than expected run-lengths of the furnaces. This was focused upon and addressed.

In Case example B, it is seen that the production is at 90% of the design level, on account of yield being at 82% of its design level and partly because of the availability being at 95% of design. Further drill down in this case showed that the yield gap was on account of an anomalous operation of a few furnaces, caused by instrument errors driving the APC incorrectly. Some of the yield gap in this case was also found to arise from losses in the recovery section of the plant. The Availability gap was again on account of furnace run-lengths.

The OAE helps us focus on the area to work on that impacts the bigger picture goal. As we observed above, in the Case Example A, the immediate area to focus on was the availability, since the plant seems to be doing well with respect to throughput and yield, atleast versus design, even though it is possible for them to be doing better on these parameters (when comparing with ideal possible) and they have done better before on these parameters. In Case Example B, Yield was the immediate focus area, followed by availability as a first step, while the plant could be doing better on all three prameters and hence production, versus what is ideally possible.

The OAE is a high level metric, which drills down to more detailed analogous defined metrics (OEE) at the section or equipment level, to identify the root case of the gaps and allowing focus on the area where the fix is required.

Key Challenges to be addressed

One of the key challenges that make it difficult to use the measured and stored data from the manufacturing process, as is, are Errors in data. Recent surveys have consistently shown this to be the top challenge[1], organizations face when trying to use the collected and archived data, as is.

[1] *Business Intelligence, Analytics and data management survey in late 2015 by TDWI showed data errors to be the number one challenge (@40%), followed by analytics skills (@38%) (further details covered in Reference 1).*

The data errors results in non-closure of the material balance around the plant.

The OAE metric, and the visual summary provided in Figures 1 and 2 cannot be meaningfully calculated with raw data. The data errors could give an erroneous picture of the relationship and gaps beteween the parameters. The ideal benchmark has to be calculated based on boundary conditions around the plant, like feed and product specs and offtake as well as equipment well being and status. The numbers for design and historical best have to be picked depending on the feed slate, the furnace configuration, etc., so we are comparing "apples to apples".

Technologies now available can be leveraged to address the data error and dynamic benchmark challenges.

Paradigm shift to leverage Industry 4.0 technologies to improve OAE

There has been a lot of media coverage on Industry 4.0. There is school of thought that believes it to be the fourth Industrial Revolution in the making (Figure 3a). Industry 4.0 is the cyber-physical transformation of manufacturing that promotes connected manufacturing and a digital convergence between industry, businesses and other processes.

The vision of Industry 4.0 will result in a smart factory and make full use of digital technologies for manufacturing. It is currently in its infacncy and beginning to take place in isolated ways. It is expected to be transformative, with end-to-end digitalization, resulting in a fully interoperable ecosystem of machines and partners across the supply chain, with data both informing and correcting the course of action.

Figure 3a

The Industrial Revolutions and where Industry 4.0 fits in historically

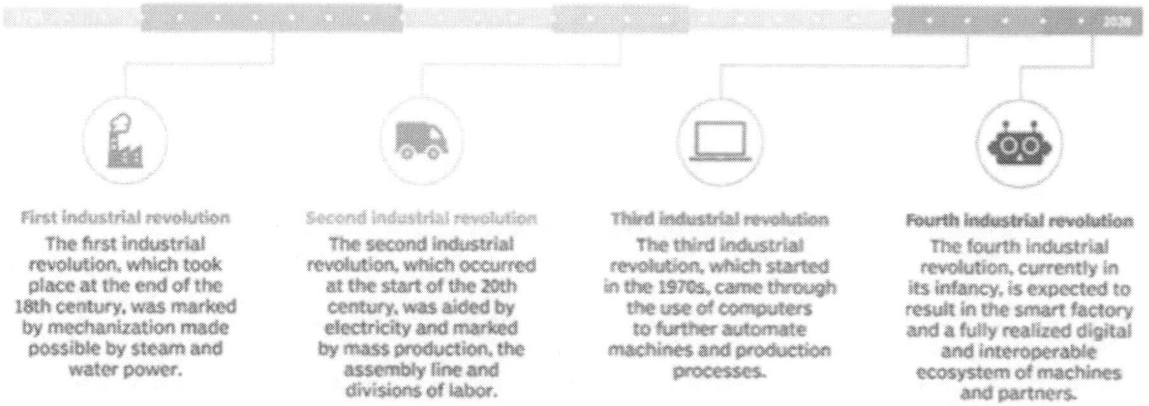

Source: Definition Industry 4.0, Tech Target, March 26, 2018

Industry 4.0 has its basis in Digital assets (technologies and concepts). Some of the technologies/concepts that are behind the Industry 4.0 discussion are:

- **Big Data Analytics**
- **Digital Twin**
- **Machine Learning**
- Cloud
- **Predictive Analytics**
- **IoT**
- **AI**
- OPC UA
- **Adaptive Analytics**
- **Everything-as-a-service**
- **Smart Sensors**
- Data Lakes

While all of the technologies may not have direct relevance to the process manufacturing application, right now, combining the applicable ones in the right way can add tremendous value to improve operational efficiency. Figure 3b below shows that within the digital asset life cycle management category, Operations performance effectiveness improvement has a significant value for the industry. The key is to leverage these technologies, without getting too far ahead on implementing bleeding edge techniques versus leading edge and feasible ones.

Figure 3b

The Significance of Operations Performance Improvement in Digitization

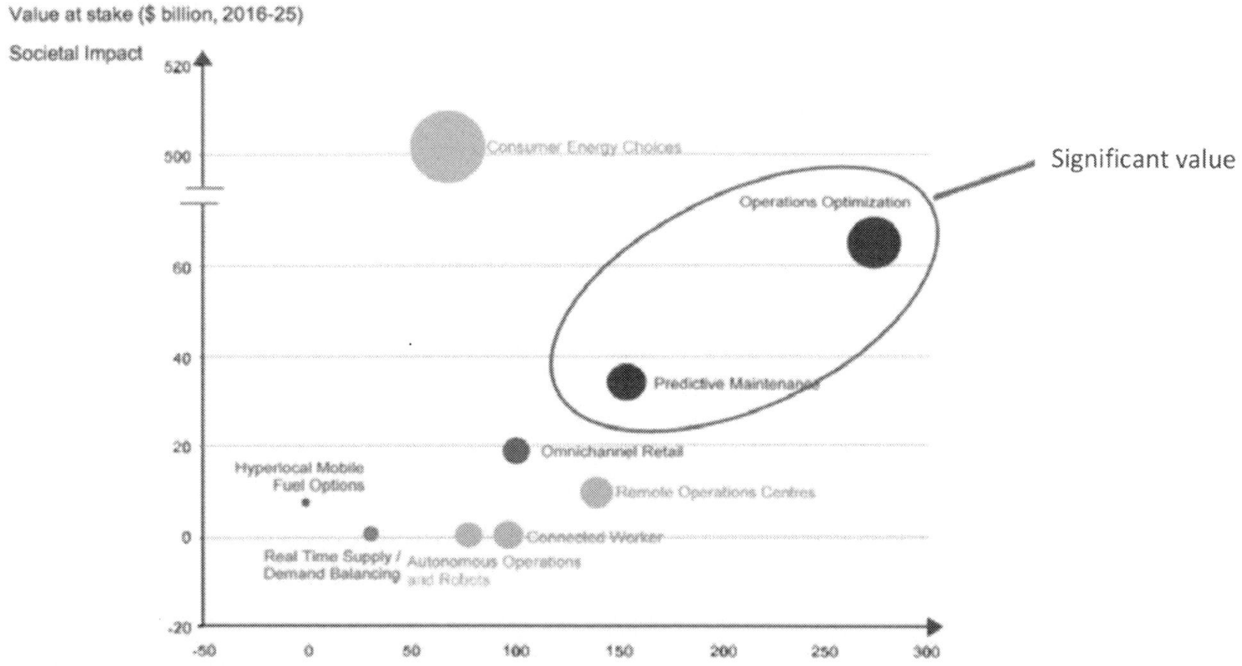

Source: World Economic Forum/Accenture analysis

Industry impact $ billion

Table 1 below shows paradigms versus possible improvements leveraging Industry 4.0.

Table 1

Evolution of Managing Process Operations – Paradigm changes

<table>
<tr><td rowspan="2">Paradigm
(since 1990's)</td><td colspan="2">Industry 4.0
- ></td></tr>
<tr><td>Current possible
(Digitalized application)</td><td>Future
(Automated applications)</td></tr>
<tr><td>• Machine automation
• Advanced Process Control (DMC)
• Onsite operations support by process engineers
• Access to data (but overload)
• Access to modeling software (but typically unsused)
• Operator/engineer personal experience and intuition based operation
• Guided by licensor SOPs</td><td>• Big Data analytics
• Digital Twin
o Predictive
o Adaptive
• Collaboration between Digital asset experts + process engineer team
• IoT
o Collaborative remote operations engineering
o Operations decision support as a service
• Interactive</td><td>• Automate using AI
• Auto adaptive analytics
• Reduce human intervention in the analytics process</td></tr>
<tr><td colspan="3">Outcome</td></tr>
<tr><td>• Early detection of anomalies often missed due to inability to uncover truths and insights that aren't readily obvious or don't align with conventional process wisdom or intuition</td><td>• Insights not otherwise captured
• Early attention to anomalies and actions to be taken that may otherwise slip attention
• Improved efficiency
• Improved Reliability</td><td>• Faster realization time and action</td></tr>
</table>

Companies that leverage digital technologies to simplify, automate and optimize operations, better than paradigms allow, will be at the forefront of transforming the industry to ensure they thrive in the volatile market conditons. Big Data Analytics is one of the technologies that is often discussed under Industry 4.0. Like many of the other technologies, stand alone Big Data Analytics may not deliver the expected value. It has to be combined with other techniques to be able to extract value for operations. As a result, the boundaries between some of these technologies often seem blurred. An effective utilization of Big Data Analytics entails:

- The development and adaptive maintenance of a Digital Twin
- The collaborative working between process engineers, data scientists and/or digital asset experts
- Effective use of IoT to leverage remote experts by taking the data to the experts, instead of the other way around to address the skills challenge and also lend focus
- The thought of using the analytics support-as-a-service
- Dashboard for effective visualization of information generated by the analysis

An overview of the Digitalized system for process manufacturing is shown in Figure 4 below.

Figure 4

Overview of Digitalized system for Process Manufacturing

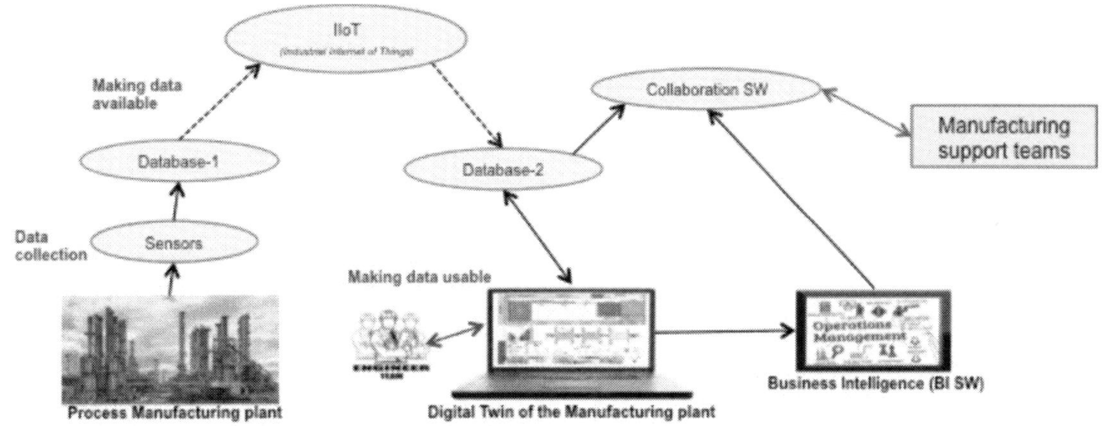

Big Data Analytics and the Digital Twin in Process Manufacturing

Data availability is not a contstraint in most Ethylene plants and process manufacturing plants in general. Plant data is collected at regular intervals in the order of seconds from the control sysytem and made available on plant historians. In additon Laboratory Information systems, operator logs provide additional, albeit less frequent data. Additional information about the process configuration is available in engineering drawings and data sheets. Business data that drives plant targets is available from the ERP system. There is no dearth of data, in fact there is an overload. Utilization of data is the constraint. As shown in Figure 5 below, more than 90% of the data collected is archived and not even looked at. The data available from a plant qualifies as Big Data.

Figure 5

Big Data in Process Manufacturing – limited use paradigm

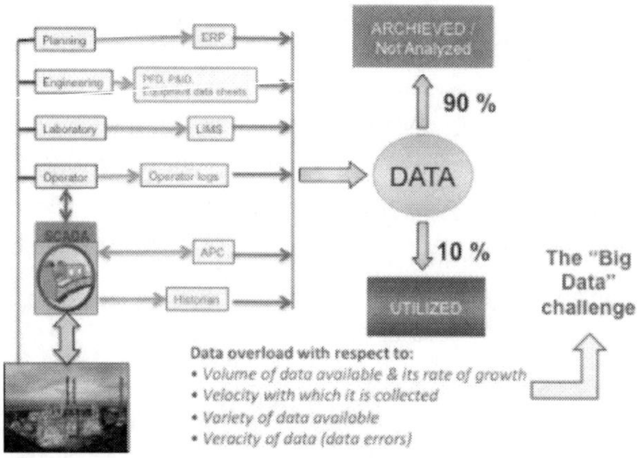

Big Data Analytics has been widely used of late in Social Media and Customer/Market analysis. However, Big Data Analytics in Process Manufacturing has to be handled differently from its current popular applications in other fields.

- In Process Analytics the "Why?" is as important as the "What?" as a result from the Analytics, unlike in Media and Customer analytics, where the "What?" is largely sufficient
  - ✓ Find and interpret patterns in data
  - ✓ Use for predictive purposes
  - ✓ Extract meaningful relationships to use to improve the process
- In Process Big Data Analytics, need combine Machine Learning models with Fundamental models
  - ✓ Machine Learning models
    - ➢ Predictive
  - ✓ Fundamental models
    - ➢ Surrogate models to explain
    - ➢ Predictive in a larger operating range

The Analysis engine for analyzing the big data from process manufacturing has to be a combination of Machine Learning models and fundamental models. A more detailed coverage of the application of Big Data Analytics to Process Manufacturing may be found in Reference 1. A case example of the type of models that go behind the Ethylene process manufacturing analysis and the benefits from its use, is shown in Figure 6. The Machine Learning models are driven purely by the data collected from the plant. The fundamental models help explain the physics and chemistry and the why, behind the observations from the data driven models.

Figure 6

Process Manufacturing Big Data Analytics requires a combination of models

Models and Analysis *(combination of models used)*

- Interactive dashboard
- Special Studies
- DMC Variable Tracker
- Coking Model: Furnace Run-length Predictor
- Machine learning and statistical model
- Performance Benchmarking
- Overall equipment Effectiveness Metric
- Equipment Health Monitoring (EHM)
- Safety Deviation Tracker
- Yield Model (COILSIM1D)
- Maintained Steady-state model (Hysys)
- Convection section model (HTRI)
- Exchanger HTRI models

Benefits

- Maximizing 13 furnace operation
- Improving furnace run lengths
- Optimizing furnace performance (i.e. conversion and yield with better health)
- Improved reliability
- Extending time between turnarounds
- Better control and less variability
- Cold section capacity increase

This basket of models integrated with one another, provides an accurate representation of the asset (the ethylene plant). Maintenance of these models is necessary to keep it mirroring the plant conditions as it continuously changes over time, on account of changing feed conditions, environmental conditions, equipment condition and product requirement, so that it can mimic the behaviour of the plant.

This system of models can predict how the plant will behave to changing conditions, it can explain the root cause of certain observations, it can help identify errors in measurements and it is a useful tool to run scenarios and what-ifs, to see how the plant will behave, without actually testing it on the plant.

In the Industry 4.0 parlance, this is referred to as the Digital twin, a term introduced by GE originally, albeit as a support for design. Gartner has included the Digital Twin in the top five digitization trends in 2017 and expects several things to be represented by digital twins over the next five years. It has been defined as "*a dynamic software model of a physical thing or system, that can be used to analyze and simulate real world conditions, response to changes, improve operations and add value, based on data provided by sensors in the physical world and the physics of how the components of a thing operate and respond to the environment*".

The utility of the digital twin in product design has been well appreciated. Its utility in process manufacturing is also very significant as was elaborated in detail in the Reference 1 paper. A significant utility of having a maintained digital twin (a compendium of tuned models) is required to address the significant data errors challenge as explained in greater detail in the paper in Reference 1. However, it is important to understand, what it takes to develop and maintain it and the processes that have to be structured around it to make it useful and deliver value. Developing a model or a set of models to mirror an ongoing operation and maintaining it is a far more complex exerecise than developing and using a design quality model. However, the benefits and value of doing this far exceeds the effort and cost.

Functionalizing the Digital Twin & Enabling Chemical Process Big Data Analytics

While the Industry 4.0 concepts and techniques can be very useful to improve performance benchmarks of process manufacturing, it requires changes to the way the operations is run and managed. Data Analytics is the base for value delivering decision making in this digitized paradigm.

Certain thoughts that are considered sacrosanct based on operating paradigms need to be overcome and changes made to based on certain realities, in order to make use of the Digital Twin and Big Data Analytics:

- The thought that by monitoring the data, insights will be gained and suggestions made to operations will generate lasting change is incorrect.
- Data availability and dashboards alone do not deliver results – sustained analysis and communication is key to knowledge and value capture
- It has to be appreciated that implementing changes to operating conditions in the plant is not automatic or easy
    - Change have to happen at the site of control of equipment and the changes require monitoring, correction, ongoing communication and convincing
    - Significant analysis is required to generate insights and actions that will consistently add value and provide the confidence to drive the change at the operations level
- Models built and maintained to reflect actual plant conditions and that can predict its behavior are essential to gain actionable insights based on analysis instead of intuitional feel and directional guidance
- Implementation requires communication with operations that is based on analytics to instill confidence and make it happen (i.e. "your thoughts are not better than my thoughts"...but if digital twin showed results from change, I can make it happen)

The typical operations analytics paradigm:

- A typical plant has an occasional converged process simulation of the plant run to check for general agreement with operation on an infrequent basis (i.e. monthly)
    - The ground work necessary to perform and sustain analysis is significant and requires a specialized engineer
        - ✓ Monthly runs planned but they rarely happen
    - Substantial effort is required with data and model preparation, maintenance and use, to get distilled decision quality information

        - ✓ It requires application with expertise and persistence on a frequent basis to keep it pertinent for regular use
        - ✓ An unmaintained model may provide "directional guidance", at best, because it is unable to get to decision quality specifics and answer questions and often not offer the confidence in the results
        - ✓ The process simulation by itself does not characterize the end to end ethylene plant operation, but only portions of it
    - Engineers typically spend ~80% of their time finding and validating information versus running and analyzing result
        - ✓ As a result, little model based analytics ever get done
- The work and frequent nature of using the tuned models that mirror the plant is what results in the value, which is not the typical paradigm in the production and technical support environment

Even if data capture, storage and dashboard presentation is good, utilization requires knowing:

- How to clean and filter that data for use, identifying and weeding out bad data
    - Veracity of the collected data is a major challenge that restricts its use
- The analytics tools required for the desired result
    - The analytics and tools to apply requires knowledge of techniques as well as the manufacturing process (the parts of the digital twin engine)
- How to manage the analytics tools behind the digital twin
    - To build models to reflect reality and tweaking to ensure convergence
    - To manage the typical convergence issues of analytics tools and models
    - To go beyond just data driven analysis and use fundamental models required for process insights

A unique combination of expertise (digital asset experts) is required to address these issues that typically hinder utilization. This is often most efficiently done by having this done as a service, by an independent remote team, which allows the operating company teams to focus on prioritizing actions and getting them implemented instead of getting bogged down getting to a base of actionable information from which results can be delivered. The IoT concept makes this possible by having data taken to the relevant remote experts instead of having these experts come to the data, which is often not feasible and sustainable. This application of IoT addresses the skill availability challenge identified in the TDWI survey.

Figure 7 shows this division of effort and collaboration and the reality of keeping the digital twin functional and of ensuring conversion of data to usable and actionable information. The Manufacturer operations support team's productivity improves by being able to focus on the necessary and value added tasks, of understanding, prioritizing, ensuring implementation and thinking beyond instead of spending time on validating information and maintaining the digital twin.

Over time, several of these tasks will be automated using AI techniques (a combination of Machine Learning and validated rule based programs). While the AI algorithms and techniques to do this are available today, configuring these is more time consuming and requires the process shown in Figure 7 to have functioned for some time to capture the "learnings" and patterns from the work process and replace repeatable and predictable learning parts of it using the AI algorithms. These need to be rigorously tested and compared to a working system, before it can be implemented to go live.

The digital twin and this collaborative interaction between the Manufacturing operations support teams and the digital asset expert team (the digital twin is the key digital asset), as shown in Figure 8, offer a paradigm shift from typical current operation. Figure 9 describes the synergy between the various manufacturing support teams and the productivity benefits from the distributed collaborative operations engineering using the digital twin. Figure 10 shows the typical delivery from the decision support analytics-as-a-service center, that supports the digital twin.

Figure 7

Reality of keeping the Digital Twin alive – Distributed roles

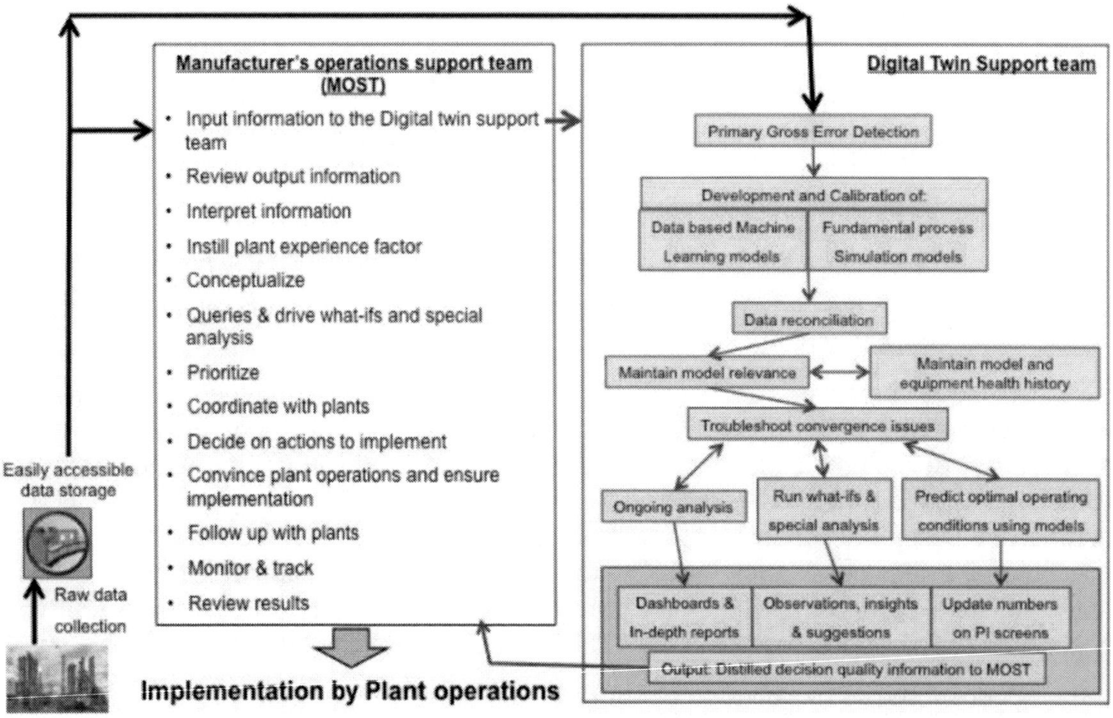

Figure 8

Collaborative operations engineering

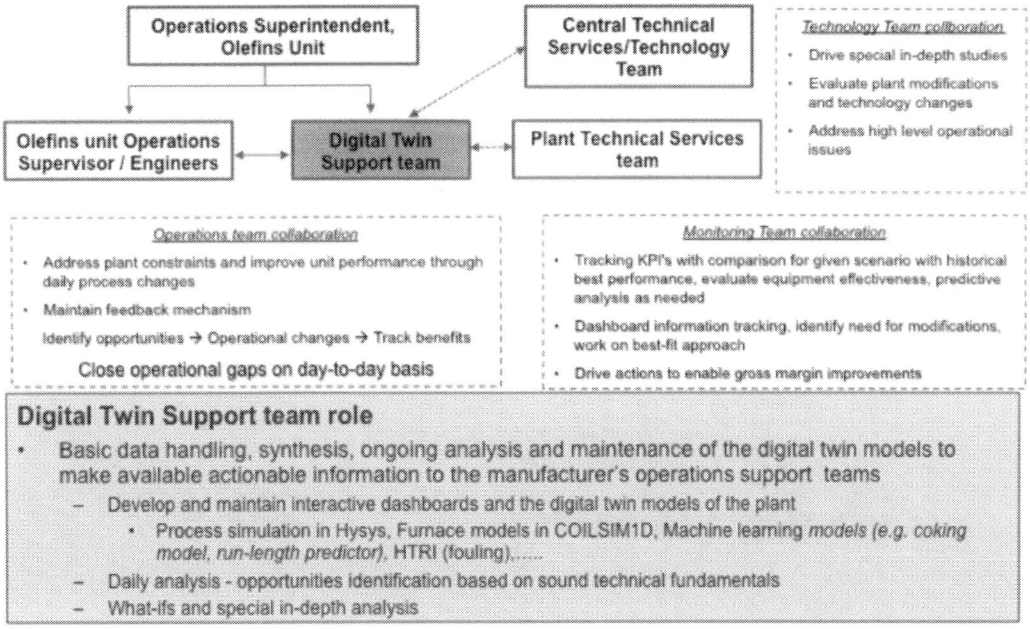

Figure 9

Facilitation of value added work by Operations support teams by the Digital Twin support team

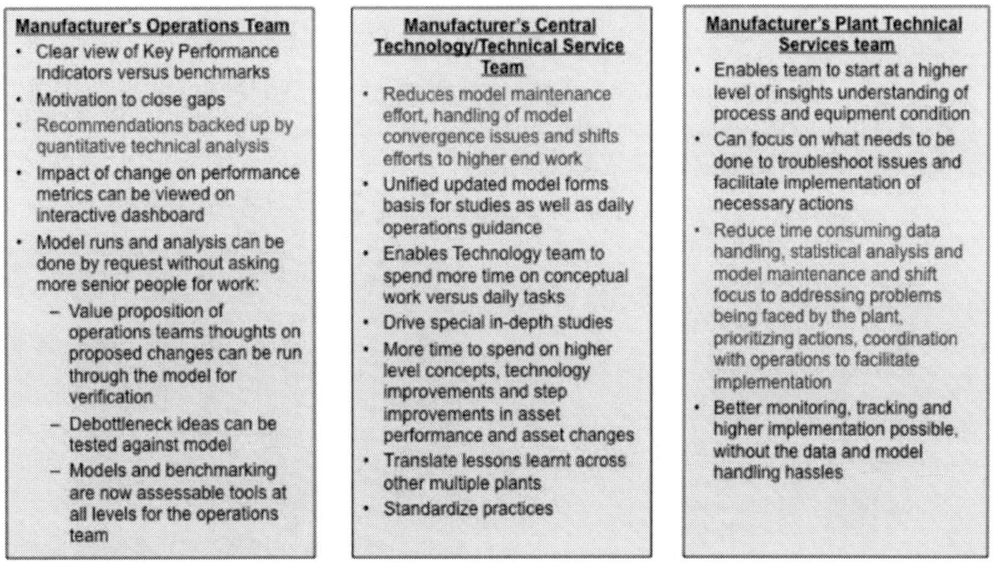

Figure 10

Typical delivery between the Collaborative Operations Engineering centers

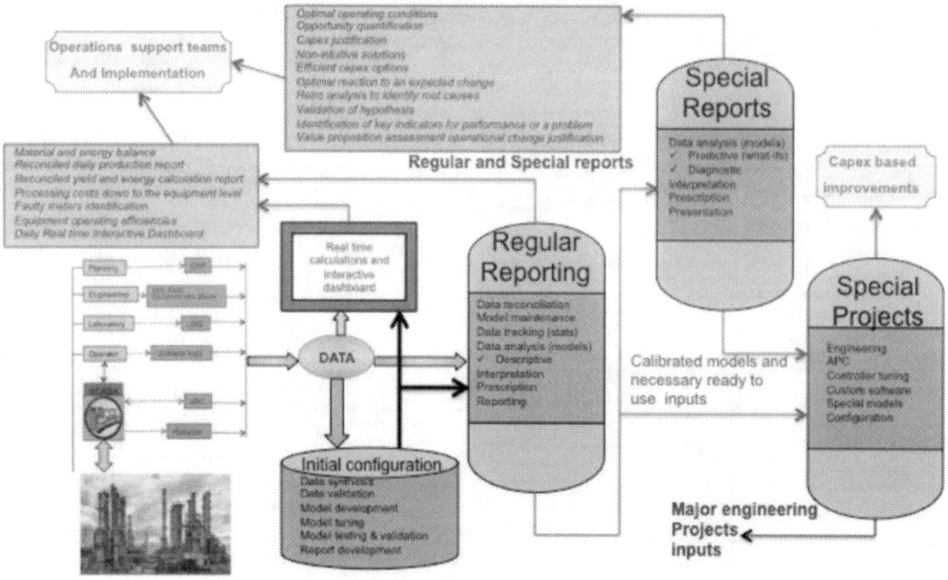

The future automated operation with AI, will utilize rules based on experience with current activity suporting the digital twin, simplified models where possible and machine learning to partially automate the digital twin support activities. The scenario over the next couple of years will likely be as shown in Figure 11, versus the current manned work process as shown in Figure 7.

Figure 11

The future scenario on operation of the Digital Twin

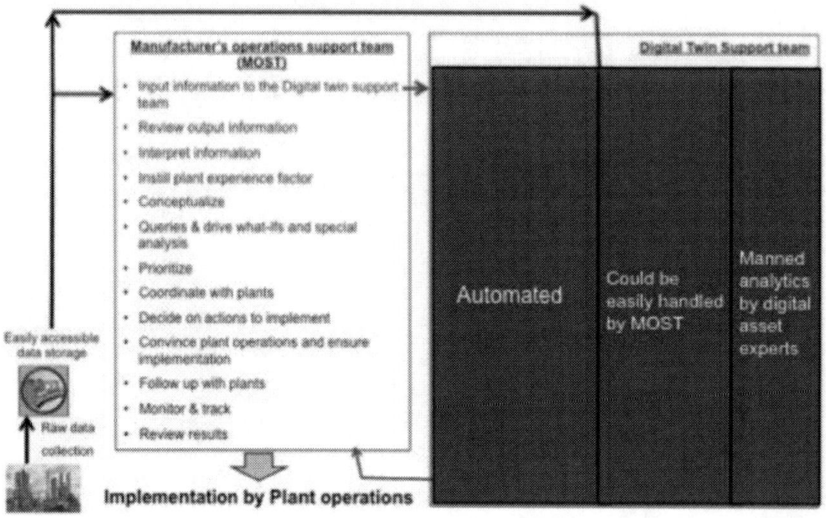

Case examples of results from application of Big Data Analysis using a Digital Twin at multiple ethylene plants

This Integrated analytics program was implemented at three ethylene sites of a major ethylene producer and has shown consistent value addition year after year. Table 2 shows a summary of some of the value derived.

Table 2

Results for three plants for a Major Ethylene Producer

Parameter	Facility A	Facility B	Facility C
Furnace Runlength	**2 x Design**	**+100%**	**+50%**
Yield	Increased	Increased	+2.3%
CGC Performance	+60 tpd	RCA solved problem	Improved
Capacity Increase	Debottleneck study	Best ever (@107%)	Max capability
Other	C2 Splitter optimized / Exchanger improve	Same + RCA	Decoke and C3 splitter improve
Significant What-if Results	25 proactive	8 proactive + 31 queries response	Numerous

The increased run-lengths helped improve the Availability Effectiveness of these assets and hence the OAE, in each of the above cases, besides lowering the costs due to fewer decokes.

Results of performing such analytics at another ethylene facility in North America are shown in Table 3.

Table 3

Improvement over previous year with Big Data Analysis using a Digital Twin

Parameter	Improvement	Remarks
Production		
Ethylene	6.8%	In spite of lighter feed slate operation
Propylene	10.1%	Record production
Furnaces overall		
Runlength	4.1%	Overall improvement
Availability	6.4%	
Variability	-1.8%	
C2" Fractionator		
Feed	5.2%	Capacity increased in spite of increased C2" in feed
C2/C2" ratio	7.1%	
Column ΔP	5.3%	
C2 spec	2.4%	

Key Success Factors – Lessons learned

Some of the key lessons learnt from the application of big data analytics to ethylene manufacturing, using the digital twin. This application has been evolved over time and with experience from its implementation and success at several ethylene sites.

- It is not just about the technology or tools. The process around the tools and application is equally, if not more important for success.
- The process and tools can be modified over time, to incorporate new technologies once they are mature enough to apply, to add efficiency to the existing work flow process, if it is able to deliver better results.
- Change management: Openness to new ways of doing things, made possible by technology. Sticking to the old process may not allow the new technology application to deliver results. For example, concepts like collaborative operations engineering, remote operations decision support-as-a-service are paradigm shifts for most operations.
- Top executive level buy-in and support is required for this application to be implemented successfully.

References:

1) P. Nair, (March 7, 2017) "Tailoring Big Data Analytics for Optimizing Ethylene Plant Asset Performance" 2017 Spring National AICHE Meeting, Paper 2B (Ethylene Producers Conference).
2) Cliff Saran, (Oct 26, 2017) "Digital Twins: Revolutionizing product business" Computer Weekly.
3) Diann Daniel, (Mar 26, 2018) "What is Indutry 4.0" Tech Target posting by Margaret Rouse.
4) L.F.Colegrove, M.B. Seasholtz, C. Khare., (March 2016) "Big Data – Getting Started on the Journey", CEP, pp. 41-45
5) Guide, (Feb 21, 2018) "Jargon Buster Guide to Artificial Intelligence" Computer Weekly.
6) P. Nair, (March 2007) "Large amounts of data requires an outside management system", Houston Business Journal.

Annexure 1

Overall Asset Effectiveness

$$P_a = T_a \times Y_a \times A_a \times \psi_{os} \times t$$
$$P_d = T_d \times Y_d \times A_d \times \psi_{os} \times t$$

$$P_a/P_d = (T_a/T_d) \times (Y_a/Y_d) \times (A_a/A_d)$$
$$OAE_a = TE_a \times YE_a \times AE_a$$

Where

P_a = Actual production (product e.g. C2", C2"+C3")
P_{hb} = Historical best production (product e.g. C2", C2"+C3")
P_I = Ideal production (product e.g. C2", C2"+C3")
P_d = Design production (product e.g. C2", C2"+C3")

T_a = Actual Feed Throughput (total feed to plant, e.g. C2, C2+C3)
T_{hb} = Historical best Feed Throughput (total feed to plant, e.g. C2, C2+C3)
T_I = Ideal Feed Throughput (total feed to plant, e.g. C2, C2+C3)
T_d = Design Feed Throughput (total feed to plant, e.g. C2, C2+C3)

Y_a = Overall actual yield of product (e.g. kgs C2"/kgs C2 fed)
Y_{hb} = Overall historical best yield of product (e.g. kgs C2"/kgs C2 fed)
Y_I = Overall ideal yield of product (e.g. kgs C2"/kgs C2 fed)
Y_d = Overall design yield of product (e.g. kgs C2"/kgs C2 fed)

A_a = Actual Availability of the Asset (the plant)
A_{hb} = Historical best Availability of the Asset (the plant)
A_I = Ideal Availability of the Asset (the plant)
A_d = Design Availability of the Asset (the plant)

ψ_{os} = On stream factor (function of time period being considered)
t = elapsed time

OAE_a = Actual Overall Asset Effectiveness = P_a/P_d
TE_a = Actual Throughput Effectiveness = T_a/T_d
YE_a = Actual Yield Effectiveness = Y_a/Y_d
AE_a = Actual Availability Effectiveness = A_a/A_d

Benchmarks:

$$OAE_{hb} = TE_{hb} \times YE_{hb} \times AE_{hb}$$
$$OAE_I = TE_I \times YE_I \times AE_I$$

OAE_{hb} = Historical best OAE = P_{hb} / P_d
OAE_I = Ideal OAE = P_I / P_d

TE_{hb}= Historical best Throughput effectiveness = T_{hb} / T_d
TE_I = Ideal Throughput effectiveness = T_I / T_d

YE_{hb}= Historical best Yield effectiveness = Y_{hb} / Y_d
YE_I = Ideal Yield effectiveness = Y_I / Y_d

AE_{hb}= Historical best Availability effectiveness = A_{hb} / A_d
AE_I = Ideal Availability effectiveness = A_I / A_d

Note:

Decoupling Availability & Throughput/Yield

- Availability at any point in time for any equipment is 0 or 1 (efficiency of working of the equipment is not reflected in the availability – this is reflected in the throughput)
- Throughput linked by downstream constraints of equipment and its efficiency (this enables delinking the Throughput and Availability factors)
- Throughput only considered over the time when the plant or equipment is available
- Availabilty of the plant = Min (availability of equipment in series and avg of equipment in parallel in the process flow)

Figure 12

Simple illustration: Concept of decoupling Throughput from Availability in the OAE calculations

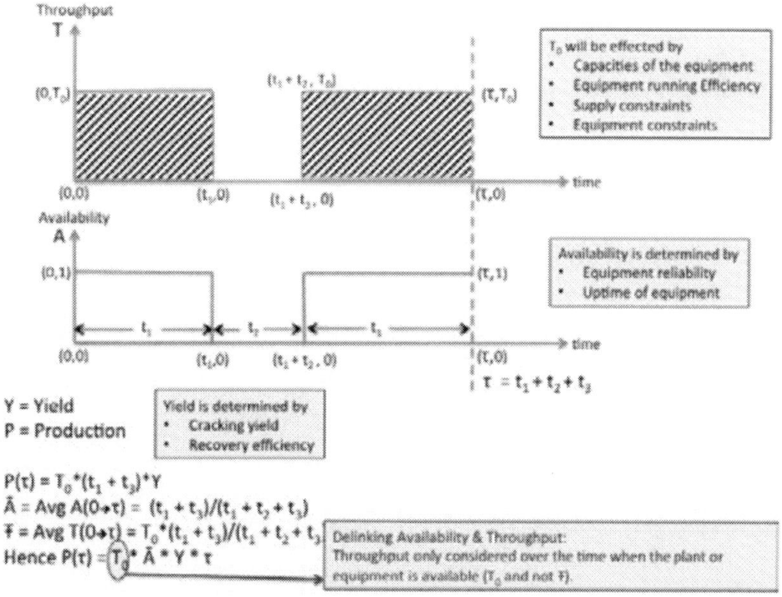

OAE Gap analysis

Gap between Design and Actual

Production gap = $1 - (P_a / P_d)$

Contribution of Throughput gap to the Production gap:
$(1 - (P_a /P_d))*(1 - (T_a/T_d))/(3 - ((T_a/T_d) + (Y_a/Y_d) + (A_a/A_d)))$

Contribution of Yield gap to the Production gap:
$(1 - (P_a /P_d))*(1 - (Y_a/Y_d))/(3 - ((T_a/T_d) + (Y_a/Y_d) + (A_a/A_d)))$

Contribution of Availability gap to the Production gap:
$(1 - (P_a /P_d))*(1 - (A_a/A_d))/(3 - ((T_a/T_d) + (Y_a/Y_d) + (A_a/A_d)))$

Gap between Historical Best and Actual

Production gap = $(P_{hb} /P_d) - (P_a /P_d)$

Contribution of Throughput gap to the Production gap:
$((P_{hb} /P_d) - (P_a /P_d))*((T_{hb}/T_d) - (T_a/T_d))/(((T_{hb}/T_d) + (Y_{hb}/Y_d) + (A_{hb}/A_d)) - ((T_a/T_d) + (Y_a/Y_d) + (A_a/A_d)))$

Contribution of Yield gap to the Production gap:
$((P_{hb} /P_d) - (P_a /P_d))*((Y_{hb}/Y_d) - (Y_a/Y_d))/(((T_{hb}/T_d) + (Y_{hb}/Y_d) + (A_{hb}/A_d)) - ((T_a/T_d) + (Y_a/Y_d) + (A_a/A_d)))$

Contribution of Availability gap to the Production gap:
$((P_{hb} /P_d) - (P_a /P_d))*((A_{hb}/A_d) - (A_a/A_d))/(((T_{hb}/T_d) + (Y_{hb}/Y_d) + (A_{hb}/A_d)) - ((T_a/T_d) + (Y_a/Y_d) + (A_a/A_d)))$

Gap between Ideal and Actual

Production gap = $(P_I /P_d) - (P_a /P_d)$

Contribution of Throughput gap to the Production gap:
$((P_I /P_d) - (P_a /P_d))*((T_I/T_d) - (T_a/T_d))/(((T_I/T_d) + (Y_I/Y_d) + (A_I/A_d)) - ((T_a/T_d) + (Y_a/Y_d) + (A_a/A_d)))$

Contribution of Yield gap to the Production gap:
$((P_I /P_d) - (P_a /P_d))*((Y_I/Y_d) - (Y_a/Y_d))/(((T_I/T_d) + (Y_I/Y_d) + (A_I/A_d)) - ((T_a/T_d) + (Y_a/Y_d) + (A_a/A_d)))$

Contribution of Availability gap to the Production gap:
$((P_I /P_d) - (P_a /P_d))*((A_I/A_d) - (A_a/A_d))/(((T_I/T_d) + (Y_I/Y_d) + (A_I/A_d)) - ((T_a/T_d) + (Y_a/Y_d) + (A_a/A_d)))$

Table 4: Ready reckoner table for OAE gap analysis calculations

Compare between	Multiplier	Numerator			Denominator
	P	T	Y	A	
Actual vs Design	$1 - (P_a/P_d)$	$1 - (T_a/T_d)$	$1 - (Y_a/Y_d)$	$1 - (A_a/A_d)$	$3 - ((T_a/T_d) + (Y_a/Y_d) + (A_a/A_d))$
Actual vs historical best	$(P_{hb}/P_d) - (P_a/P_d)$	$(T_{hb}/T_d) - (T_a/T_d)$	$(Y_{hb}/Y_d) - (Y_a/Y_d)$	$(A_{hb}/A_d) - (A_a/A_d)$	$((T_{hb}/T_d) + (Y_{hb}/Y_d) + (A_{hb}/A_d)) - ((T_a/T_d) + (Y_a/Y_d) + (A_a/A_d))$
Actual vs Ideal	$(P_I/P_d) - (P_a/P_d)$	$(T_I/T_d) - (T_a/T_d)$	$(Y_I/Y_d) - (Y_a/Y_d)$	$(A_I/A_d) - (A_a/A_d)$	$((T_I/T_d) + (Y_I/Y_d) + (A_I/A_d)) - ((T_a/T_d) + (Y_a/Y_d) + (A_a/A_d))$

AIChE Paper Number 169f

Start Listening to Your Heat Exchanger - Smart Data Application Enables Insight into Process Equipment

Dr. Jörg Weidenfeller
Director RnD / Technology
Axel Kindgen
Director IT / Digitalisation
Karsten Stückrath
President & Managing Director
SCHMIDTSCHE SCHACK / ARVOS GmbH:

Prepared for Presentation at the 2018 Spring National Meeting
Orlando, Florida, April 22-26 2018

AIChE Paper Number 169f

Start Listening to Your Heat Exchanger - Smart Data Application Enables Insight into Process Equipment

Dr. Jörg Weidenfeller
Director RnD / Technology
Axel Kindgen
Director IT / Digitalisation
Karsten Stückrath
President & Managing Director
SCHMIDTSCHE SCHACK / ARVOS GmbH:

Abstract:

Transfer line exchangers, their operation and reliability are critical components in the production of Ethylene and the economics of the plant.

Since the focus of the operator is onto the process itself, standard sensor installations are located to monitor and control the process flows. The status and condition of the heat exchanger might be derived by measured data up- and downstream the equipment and is based on individual experience of the operator. In addition regular inspections are conducted.

Transfer line exchangers equipped with additional sensors provide smart data for the operation and maintenance team. This Industrial Internet of Things (IIoT) solution is based on proven cloud technology and is adaptable/ flexible to various customer requirements. It also makes use of machine learning algorithms in order to allow predictions for different parameters.

All those smart data will be used to generate regular operation and health reports for the process heat exchanger, which provide the customer a deeper insight into his equipment and finally enhance the equipment availability.

Introduction

SCHMIDTSCHE SCHACK | ARVOS GmbH, formerly known as ALSTOM Power Energy Recovery is designer and manufacturer (OEM) for transfer line exchangers (TLE) in Ethylene industry and waste heat recovery systems in further chemical, petrochemical and metallurgical processes. The company was founded more than 100 years ago and was the first who introduced and established steam super-heaters for steam locomotives. Up to this time the boiler of steam locomotives produced only saturated steam with an appropriate poor efficiency. The founder of SCHMIDTSCHE SCHACK was driven by improving this product, increasing the efficiency of the steam process and implementing a new technology into the existing envelope of the steam locomotive.

This company philosophy and motivation have not changed. More than 50 years ago the transfer line exchanger was introduced and established in the Ethylene industry by replacing the quench. Today further improvements of the TLE are consequently introduced by keeping the existing envelope of the hardware.

The status quo in Ethylene plants is that sensors and instrumentation are installed to monitor and control the production process and the process flows. From those measurement locations upstream and downstream the transfer line exchanger the status and condition of the hardware might be derived. The interpretation of those data strongly depends on the knowledge and experience of the individual operator. The real status and condition of the transfer line exchanger is obtained during regular shut-down of the production line and inspection of the TLE. This indirect approach has uncertainties due to

- Data are not directly collected at the TLE and do not describe TLE condition
- inspection during shut-down only provides a snap-shot with limited confidence for future condition

The transfer line exchangers, their operation and reliability are critical components in an ethylene plant with an impact on the plant economics. In the transfer line exchangers the sensible heat in the cracked gas of an ethylene furnace is converted into high pressures steam (HPS). HPS is used to drive turbines, compressors and pumps in the separation train of an ethylene plant and is most important in the crackers energy system. Any loss of steam production by failure or decreased performance due to fouling of the TLE has to be compensated by HPS import from boilers requiring additional fuel gas or fuel oil with a negative effect on plant economics. A high reliability and performance of the TLEs are most important, as utilization of the TLEs have a significant impact on the plant economics and require a high quality of the TLE, for maximum utilization and performance.

In general availability and performance are independent of each other, as a heavily fouled TLE will operate at a low performance level. Maximizing the availability and performance of the TLEs by online monitoring and predictive maintenance increases the asset utilization and will improve the plant economics significantly.

Enabling The Smart TLE

McKinsey's digital compass describes that asset utilization drives value by making the best use of the hardware. Predictive maintenance thus can drive value by decreasing the planned and unplanned asset downtime. Based on sensor readings early indications of hardware problems can be detected and timely corrections can be performed. Additionally maintenance resources can be prioritized and availability can be increased. Typically, predictive maintenance decreases the total downtime by 30 to 50%.

To support the needs of Ethylene plant operators SCHMIDTSCHE SCHACK started to further improve TLE's functionality by applying IIoT technology and keeping TLE's envelope. This requires a stepwise approach as shown in Figure 1. At the base level the TLE is equipped with additional sensors. The measurement locations are at the points of interest for the TLE but not for the process. Those sensor readings are input for further evaluation and assessment of the status and condition of the TLE. The data are acquired and stored in a independent system. This approach ensures that the Data Control System (DCS) of the plant will not be impacted or disturbed. Optional operational data can be integrated by connecting to the DCS if the operator requests this. From the sensor readings meaningful values and equipment Key Performance Indicators (KPI's) are derived and monitored. Additionally the data are used for models which are based on OEM's domain knowledge with respect to thermal and mechanical design. In addition artificial intelligence (AI) is used to discover abnormalities in operation, to forecast the reliability and performance of the TLE and to provide diagnostics and future recommendations.

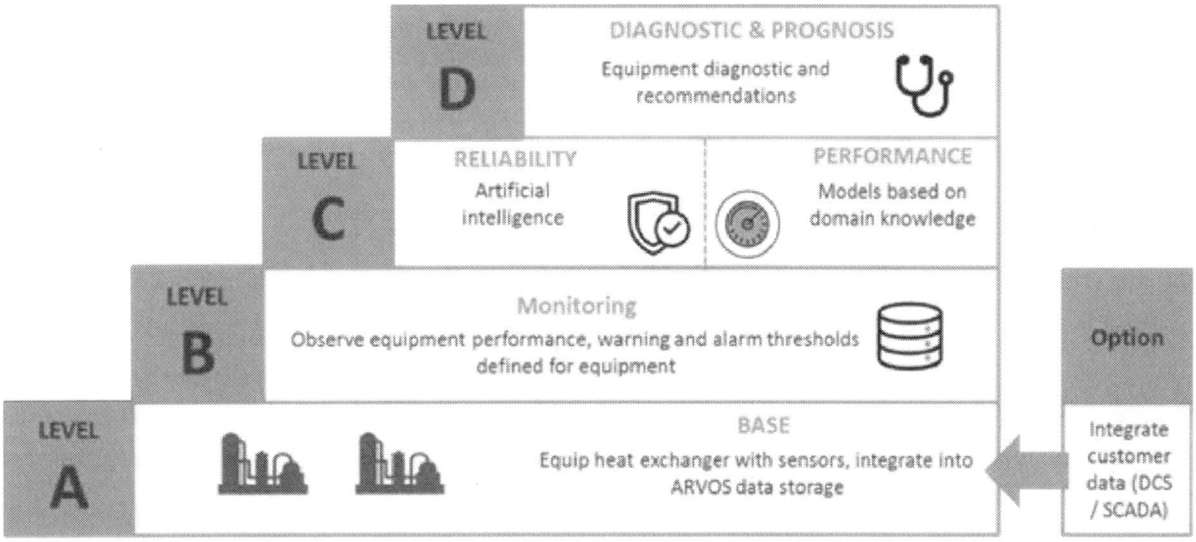

Figure 1: Stepwise approach to increase asset utilization

Technical Set-Up

The measurement locations for the additional sensors at the TLE will be installed during regular manufacturing process in the SCHMIDTSCHE SCHACK workshop. At site the sensors will be installed after erection of the TLE. The independent data acquisition system (DAQ) will also be installed nearby the TLE. The DAQ consists of Input/Output controllers connected to a edge-device which sends the live data via high-speed wireless network to the SCHMIDTSCHE

SCHACK data storage. The complete DAQ system is enclosed in an explosive protected (ATEX) control box.

The SCHMIDTSCHE SCHACK data storage manages the storage and data handling. Live data are evaluated and visualized, historical data are stored. Both types of data are used for further data manipulation by domain knowledge algorithms and machine learning. This enables diagnostic of the current TLE status, future condition prognosis and recommendation service. In summary this describes the new SCHMIDTSCHE SCHACK Zero.One platform for collection and analysis of data from transfer line exchangers, see also Figure 2. Zero.One is a cloud-based platform as a service and enables asset performance management and analytics by connecting equipment, data and people.

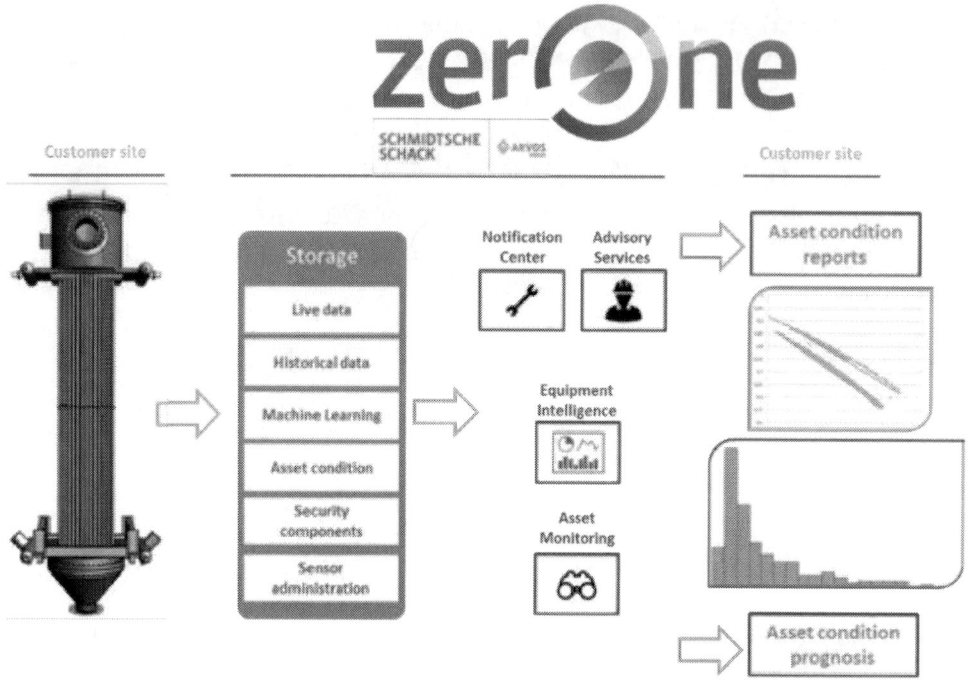

Figure 2: End to end set-up for TLE monitoring and prognosis

Operators can receive frequently condition and status reports of their TLE. Plant managers will get an overview about all TLEs in their plant. This information is available in asset, plant and fleet dashboards and reports (see Figure 3) which have different information levels.

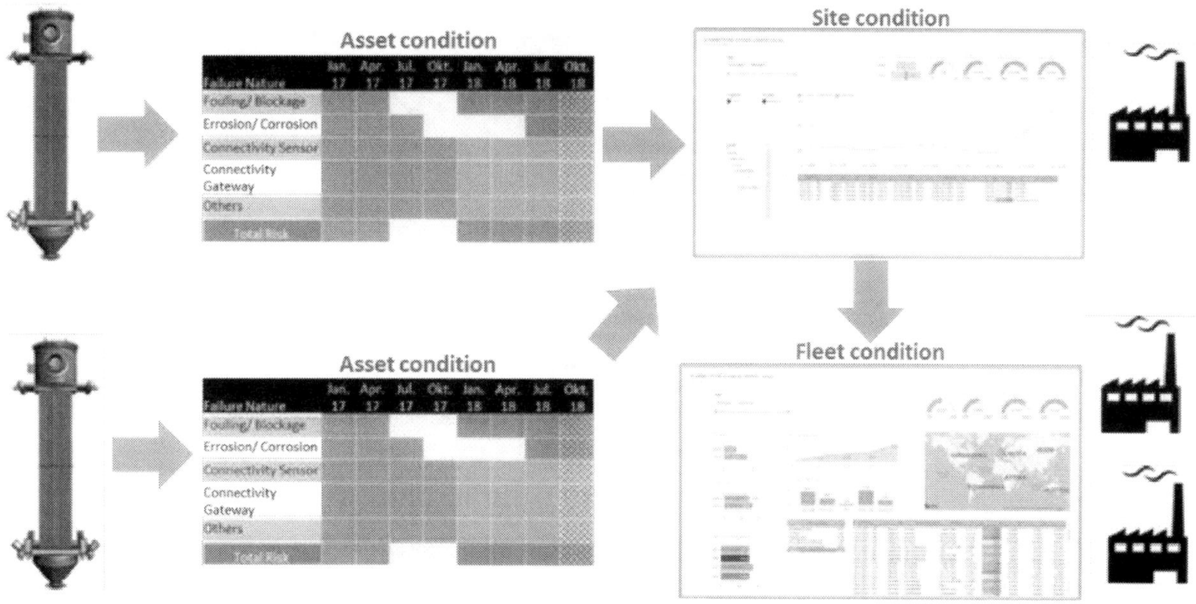

Figure 3: Dashboards for TLE condition

From Raw To Smart Data

Collecting and monitoring the operational data of a TLE, like gas inlet and outlet temperatures and pressure, water pressure etc. are an important requirement to come to a right conclusion about the equipment condition. However, only assessing the operational data is not sufficient since those do not directly reflect

- TLE design topics (thermal and mechanical),
- aging of the TLE and
- abnormalities during operation.

To address these subjects several methods for raw data post-processing are available. SCHMIDTSCHE SCHACK decided to apply an approach which combines OEM design knowledge and machine learning. In the following only examples of the current machine learning applications are discussed.

Machine Learning

To predict the future TLE performance by artificial intelligence the right approach and algorithm has to be identified. For that purpose operational data were simulated and random noise was overlaid to simulate fluctuations in the sensor readings. The principle of a machine learning algorithm is shown in Figure 4. Input values are modified by the model, which creates the predicted value. The model is trained in that way, that the error between output from the model and the target value is minimal. This will be achieved by adapting the algorithm in the model. The training of the model is finished, when the prediction achieves a pre-defined accuracy.

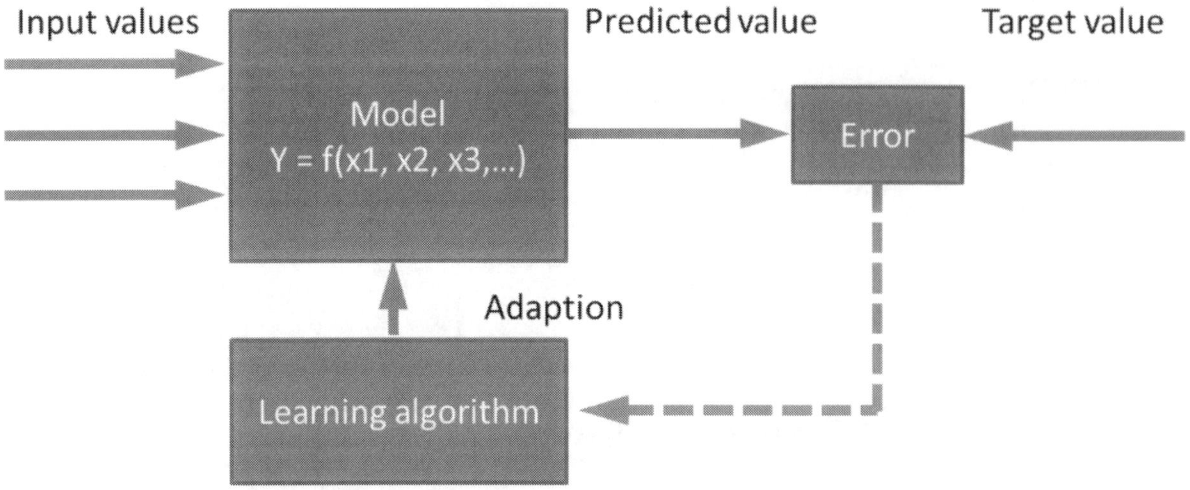

Figure 4: Principle of machine learning

When training a model, further verification is of the trained model is required to verify

- how the trained model reacts to new and unknown data
- if the training did not specialize the model in that way that accurate predictions are only possible with the training data

For that purpose a second and unknown data set has to be applied to the trained model to verify the training and the quality of prediction.

The use case for machine learning is to predict the time period to the next de-coking cycle. The signal of gas outlet temperature was used as training input. The TLE is operated first with heavy feed and after that the feed is changed to light feed until de-coking. After that the operational sequence starts again with heavy feed and again switching to light feed (see Figure 5). The training data describe 100% load and 100% heating surface.

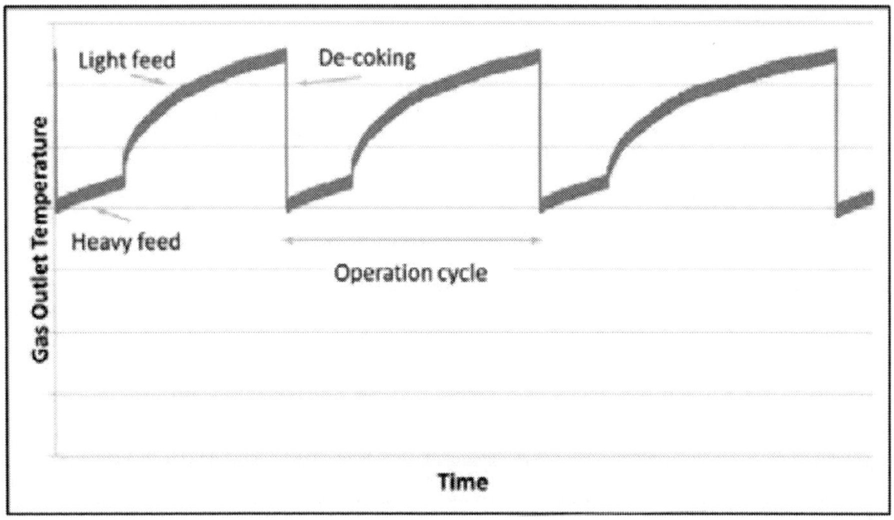

Figure 5: Training data

The training was performed only for three operation cycles. After the training the algorithm was applied to unknown test data to predict the remaining time until de-coking. In addition, several unknown boundary conditions were implemented (see Figure 6):

1. heating surface reduced by 15%, 100% load, reduced operation time
   (blockage of tubes reduces heating surface, for 100% load the gas outlet temperature increases faster, in consequence reduced operation time)
2. heating surface reduced by 15%, 90% load, normal operation time
   (operator reduces load to 90% to achieve the normal operation time)
3. no reduction of heating surface , 90% load, extended operation time
   (operator reduces load to extend operation time)
4. no reduction of heating surface , 100% load, normal operation time

The algorithm shows very good prediction with an error of ±2 days for normal light feed operation. The preceding heavy feed operation has a bigger error in prediction but the feed switch is well recognized. The algorithm also predicts the time until de-coking with a sufficient accuracy for the unknown boundary conditions of reduced heating surface, reduced load, extended or reduced operation time. In case that the training would consist also these events, a further accuracy improvement for the prediction is expected.

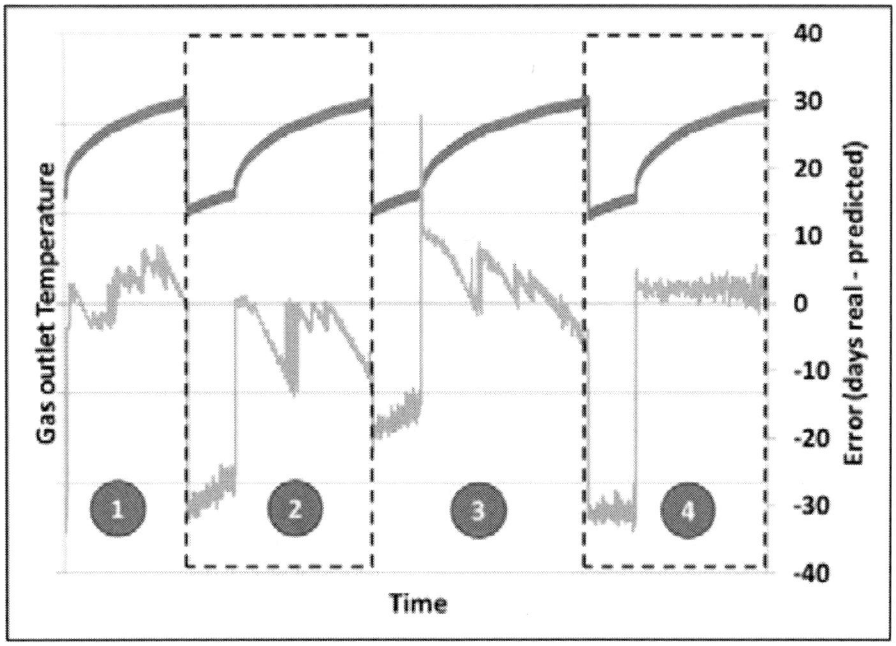

Figure 6: Test data, error between time until de-coking and prediction

Detection of Abnormality

During TLE operation several abnormalities can occur. For example single tubes can suddenly be blocked by refractory or coke particles. Blocked tubes reduce the overall heating surface of the TLE and therefore theoretically this should be visible by an increase in gas outlet temperature and pressure drop across the tubes. However, it will be difficult to detect this event by only monitoring these two signals. But with the help of artificial intelligence this will be possible.

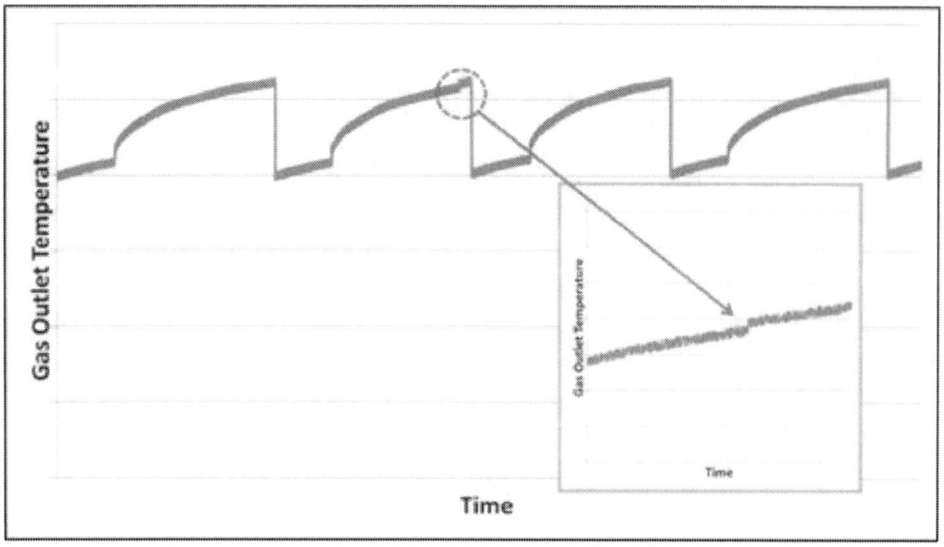

Figure 7: Temperature signal and abnormality

Figure 7 shows the gas outlet temperature which was used to train the appropriate algorithm. The small change in gas outlet temperature due to tube blockage is also shown. The algorithm was trained to detect this kind of blockage event. After the adjustment phase a second and unknown set of data was used to verify the algorithm. The result of the blockage prediction is shown in Figure 8. Here the algorithm clearly identifies the event.

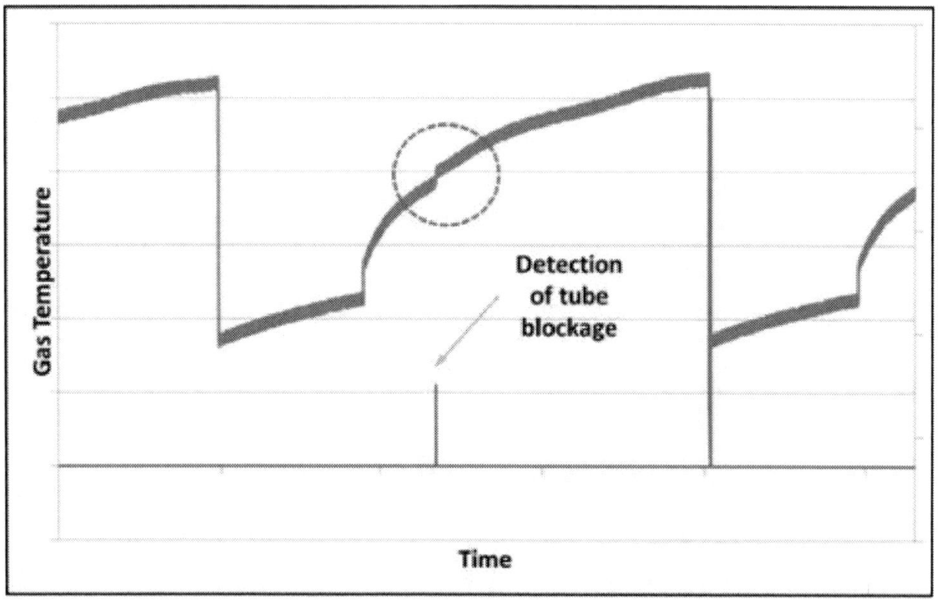

Figure 8: Tube blockage detection

Conclusion

This paper illustrates the next consequent step in further improving the functionality of SCHMIDTSCHE® transfer line exchanger by applying new technologies. Zero.One comprises additional sensors installed at the equipment and applying a cloud based platform to analyze and assess the asset. This service combines OEM design knowledge and artificial intelligence to detect abnormalities during operation and to predict future events. This will create transparency into the TLE condition, helps to optimize availability and productivity and improves planning reliability for the operator and the maintenance team.

513845 Incorporating Model-Based Technology with Big Data Concepts for Monitoring Ethylene Production

Abstract:

Data is becoming abundant and supplied in many different forms through mobile devices, wireless communications and big storage servers. With rich available data production facilities are applying "Big Data" concepts to analyze, identify and correlate parameters that indicate asset performance and health. Many model based solutions have been applied to Ethylene production providing significant benefits in monitoring process performance and optimizing production. Model based solutions feature intrinsic relationships characterized by first principle concepts or determined empirically that have been qualified for accuracy. Incorporating existing model based solutions and big data analysis offers potential for discovering process indicators that improve understanding of process and equipment performance. The resulting awareness advances process knowledge beyond "what is happening to the process" to "how did the process gets to this point" and "what other paths for the process to take" . Evaluating information generated by model based solutions and stored in large data sources can be cumbersome. Collaboration with analyzing large amounts of information and displaying discoveries in digestible form to operations and managers such as dashboards and high performance graphics often requires support from information technologist. Reliance on IT to provide a forum for data analysis and presentation of findings may restrain advances in process knowledge discoveries. Solutions that incorporate model based applications, big data analysis and provides engineers with a forum for easily analyzing information and displaying findings provides an opportunity for advancing asset performance and health.

513439 Improving Ethylene Furnace Performance with Big Data Analytics

Abstract:

Parameters around furnace performance include throughput, yield, runlength, coil life, and energy efficiency. Increasing the factors that allow maximization of any of one of these furnace performance criteria will reduce the level of performance of the other criteria. Big Data analytical techniques have been repeatedly proven to effectively find the optimal settings to meet the furnace operational objectives.

A key to successful implementation of Big Data to Ethylene furnaces is, however, to combine it with first principles reaction kinetic models. Pure empirical models work well with predicting equipment failure and thereby addressing availability issues. Availability is a variable for Ethylene furnaces to be managed. While one can argue that with sufficient data and time, empirical models can be made to replicate furnace performance over a range of feedstocks and conditions; however first principles models can also replicate performance closely with an enhanced ability to view relationships. The logical approach is to utilize the modeling approach (i.e. first principles versus empirical) that can best illustrate actual performance within the constraints of difficulty and time in building and maintaining the model. This approach can be very effectively utilized to increase production and runlength.

Adjustment variables for the furnaces include firing, draft, flow, feed balancing, and dilution steam and understanding the ultimate effect of these on coil inlet pressure, coil outlet pressure, tube metal temperature, pressure drop in TLE, firing duty, furnace efficiency, residence time, and coil outlet temperature is key to better operation of the furnace. By combining first principles modeling with predictive modeling using Big Data techniques results the ability to completely predict the performance of the furnace over the life of its run. The impact of adjustments is reflected in the performance and production of the furnace over the likewise variable runlength of the furnace. With this information furnaces can be set to optimize the production goals and targets that themselves change with feedstock variability and planned decoke scheduling.

In this presentation, models utilized will be described and their output provided. Use of this output to identify furnace anomalies at various stages in its life and the ability to take corrective actions will be shown. While dependent on feedstock availability and downstream limitations, achieving anywhere from 5% to 20% increase in production has occurred in numerous facilities.

513927 Obtaining Parsimonious Regression Models with Large Datasets

Abstract:

The advantages of using feature selection methods[1] have been long established in the literature and include lower dimensionality of the resulting dataset, easier identification of important features and lower risk of overfitting. These advantages are critical assets in today's context of big data and in the advent of Industry 4.0 initiatives, where a large and increasing number of variables and features are obtained. However, larger data sets do not always correspond to additional sources of variability and the opposite scenario is often observed: the ratio between the critical features and the irrelevant features tends to decrease to zero as the data set increases. Therefore, in this paper, a two-stage approach to develop regression models for large data sets is proposed. In the first stage, a feature selection method is proposed with the objective of removing noisy and irrelevant features, while also reducing the number of missed detections or the number of relevant variables not selected. In the second stage, the set of selected features is then combined with a regression method, which might incorporate a further step of feature selection. Two simulated datasets for continuous and batch processes, and an industrial dataset were used as case studies to validate the effectiveness of the proposed approach. Using feature selection methods lead to the selection of important predictor variables while also improving prediction performance since the models developed were more parsimonious.

Three main classes of feature selection approaches are available[2]: filter methods, wrapping methods and embedded methods. Filters are mostly based on univariate measures of association between predictors and response variables, and tend to be more efficient than wrappers and embedded methods. Analyzing the literature on filter methods, a comprehensive list of different filters has been proposed and tested for classification tasks[3], while filters for regression problems remain vastly unexplored[4]. This research focuses on mitigating this gap by assessing and comparing the performance of different filters for feature selection in regression problems. Various association metrics are considered, including Pearson's correlation coefficient, Spearman's correlation, Kendall's correlation, mutual information, and also combinations of mutual information with other filters. These filters have the flexibility to account for various relationships between predictors and response variables and can capture linear correlations (Pearson's correlation), monotonic relationships (Spearman's correlation) and non-linear associations (mutual information).

Two key performance indicators (KPI) are utilized to quantify the performance of the different filters. The first KPI measures whether relevant variables are selected and noisy ones are removed. This KPI is only used in a simulation setting because the data-generating mechanisms are known. The second KPI assesses the improvements in prediction performance obtained when different filters are used prior to model building. Furthermore, various regression methods, including Partial Least Squares (PLS) regression, were also tested in order to assess their interactions with the filters considered.

References

[1] R. Kohavi and G. H. John, "Wrappers for feature subset selection," *Artif. Intell.*, vol. 97, no. 1–2, pp. 273–324, 1997.

[2] V. Kumar and S. Minz, "Feature Selection : A literature Review," *Smart Comput. Rev.*, vol. 4, no. 3, pp. 211–229, 2014.

[3] G. Chandrashekar and F. Sahin, "A survey on feature selection methods," *Comput. Electr. Eng.*, vol. 40, no. 1, pp. 16–28, 2014.

[4] I. Guyon, A. Elisseeff, and A. M. De, "An Introduction to Variable and Feature Selection," *J. Mach. Learn. Res.*, vol. 3, pp. 1157–1182, 2003.

514251 Utilizing Deep Reinforcement Learning for Supply Chain Materials Planning

Abstract:

This paper presents a novel approach to supply chain planning optimization through the utilization of deep reinforcement learning and compares it to modern, mathematical optimization techniques. The supply chain is modeled as a factory with multiple products and stochastic demand for each product. The deep reinforcement learning agent interacts with this environment to learn by experience, as captured in neural networks, to determine which product to produce and when to produce it to satisfy customer demand. The neural networks approximate the value of each state with respect to a value function that accounts for on time delivery, the level of inventory on hand, and the customer service level. The reinforcement learning agent learns through simulations of the manufacturing environment and demand patterns in order to maximize the reward received through the value function. In parallel, we formulate a mixed-integer program that has to make identical decisions under the same constraints. The mixed-integer program can see orders out to a certain horizon, and does not use any information from historical order patterns. We compare and contrast the performance between the two methodologies on several scheduling test examples, and suggest how to choose between them, as well as how they may be used in concert.

513564 Equipment Health Monitoring to Improve Reliability and Plant Performance through Application of Data Analytics and Predictive Modeling Techniques

Abstract:

Equipment health and reliability plays a critical role in overall ethylene plant operation strategy and consistently maintaining high production rates. Enhanced equipment availability and reliability can be achieved through effective data analytics by proper interpretation, analysis and use. Knowledge acquired through data can be further put to use for managing current problems and for determining strategy for future. This paper explains how application of data analytics through modeling techniques and systematic monitoring of equipment health is effectively utilized to identify anomalies and solutions to improve equipment reliability and performance.

An **intelligent proactive** approach needs to be followed, which entails a continuous assessment of measured and derived parameters against known engineering boundaries to detect and correct problems before failures occur. Apt techniques, tools and equipment models are necessary to track the performance and identify quantum of deviation from normal. Further, predictive and prescriptive analytics helps determine the impact of deviation, need for corrective action and reach challenging solutions.

Specific Techniques for Reliability Improvement

- Statistical tools and machine learning models
    - Track and address deviations leading to equipment failures and loss of performance
- Conditional monitoring
    - Enhance the equipment reliability and availability through effective data management
- Life Estimation/Failure Prediction
    - Failure modes identification, typical life and mean time between failures etc.
- Degradation Analysis and Loss of function of equipment
    - Based on operation bands, tracking loss of function based on process parameters to flag off abnormal operation that could lead to failure
- Predictive Analysis for operations and maintenance decision making
    - Estimate the optimal time for scheduling maintenance based on the estimated loss in production while continuing to operate an unhealthy equipment

While effective multi-modeling is key, it has to be combined with the process of data collection, transforming it into information and knowledge using the models, visualization and dashboards to help interpretation and early detection and then an implementation and tracking process to ensure benefit realization. Successful application of the above process has significantly helped in improving the health of key equipment in ethylene units resulting in,

- Longer mean time between machine overhauls/ replacement à extending T/A intervals
- Improving efficiency and overall effectiveness à increasing plant production capacity depending on the severity of operating constraints

513946 Deep Reinforcement Learning for Model Predictive Control

Abstract:

We propose a deep reinforcement learning algorithm for designing process controllers. The classical approach to designing controllers involves extensive data collection, development of a mathematical model and finally, design of an appropriate controller. This approach can be rather time consuming and requires frequent maintenance of models in response to changes in process conditions. In chemical engineering, there are no good solutions to the controller design and model maintenance problem for complex systems that are nonlinear, high-dimensional and stochastic. We propose an algorithm that uses reinforcement learning to train a deep neural network for learning optimal control policies. One of the key challenges with reinforcement learning is that the learning process often requires tremendous amount of data that is, in practice, difficult to obtain from physical processes. To address that problem, our algorithm samples from a distribution of process observations and uses predictions from a model predictive controller (MPC) that is designed offline. The algorithm maps process observations to control policies and learning is guided by the MPC outputs. Guided policy search reduces training data requirements and has been demonstrated successfully for robotic control [arXiv:1504.00702] and autonomous aerial vehicles [arXiv:1509.06791]. There are several advantages to this approach for designing controllers: (1) Reduction in the size of training data. (2) An explicit model is not required. (3) An explicit control law is not required. (4) Model maintenance is handled by the reinforcement learning algorithm. (5) The algorithm works on complex, non-linear processes.

512727 Powerful and Novel Multivariate Statistical Approaches in Big Data Sets and in Data Mining with Applications to Bio-, Medical-, and Material-Informatics

Abstract:

Advanced statistical metholodogies have key roles to contribute in data mining and informatics for large data sets. Over the years, our research has developed a number of statistical techniques exploiting multivariate analysis and methodologies in many applications including bioinformatics, specifically, microarray data sets in a number of applications, medical informatics, including disesase diagnosis and discovery, and in material informatics, including the development and evaluation of material properties and testing techniques. In this talk we present the tools and methodologies that we have developed over the years and discuss their attributes and strengths. The two primary multivariate statistical methodologies that we have exploited have been principal component analysis (PCA) and cluster analysis (CA). This talk will break this techniques down for the non-expert and then demonstrate their strengths in handling large data sets to extract critical information that can be exploited in analysis, inference, diagnosis, and discovery.

514059 Image Classification in Manufacturing Analytics: Improving a Pellet Classification System with Deep Neural Networks

Abstract:

Manufacturing analytics is of paramount importance in many plants today, and its significance increases in the current context of big data analytics and industry 4.0 initiatives[1]. The fields of statistics, chemometrics, and machine learning are expected to provide tools that effectively handle data characteristics, such as process measurements (e.g. temperature, pressure), spectra from online analyzers and quality control labs, pictures taken with standard or hyper-spectral cameras, and all other data sources that provide insight into the state of product processes and quality.

The task of image classification[2] is considered in this paper. A supervised learning problem is presented whereby an image is the input and the output is a unique label attributed to the image from a set of available classes. Image classification is one of the main tools used for quality assurance and control at Dow Chemical, and developing a suitable classifier is a relevant industrial challenge due to accuracy and robustness requirements. In this context, recent developments in deep learning[3] that have proven successful in increasing image classification accuracy and providing state-of-the-art results in computer vision. Traditional approaches utilized for image classification are based on prior knowledge and pre-defined features. However, in this work, we leverage deep neural networksâ€™ (DNN) ability to automatically learn features from images[4] and test their performance in a real industrial context of pellet shape prediction. Moreover, other less complex techniques such as partial least squares discriminant analysis (PLS-DA) and random forests (RF) are explored in order to assess the benefits of adopting DNN as opposed to a more traditional classifier.

PLS-DA, RF, and DNN models were developed for two classification tasks: pellet shape classification (distinguishing good and bad pellets), and detecting tails in a pellet (distinguishing whether a pellet contains tails or not). After developing these models, the results were consistent for both classification objectives. Compared to the in situ classification system currently used at Dow Chemical, RF were able to better utilize the same pre-defined features and improve prediction accuracy significantly, while PLS-DA had the worst performance. DNN obtained the highest accuracy overall (higher than 96%), and the main advantage is that there is no need to specify a priori features because they are extracted from the raw image itself. Furthermore, visualizing the output of some layers of the network showed that activations occurred in regions that are meaningful for the classification tasks, further supporting that DNN effectively modeled the relevant features of the pellet.

References

1. Reis, M. and G. Gins, *Industrial Process Monitoring in the Big Data/Industry 4.0 Era: from Detection, to Diagnosis, to Prognosis.* Processes, 2017. **5**(3): p. 35.
2. Duchesne, C., J.J. Liu, and J.F. MacGregor, *Multivariate image analysis in the process industries: A review.* Chemometrics and Intelligent Laboratory Systems, 2012. **117**(Supplement C): p. 116-128.
3. LeCun, Y., Y. Bengio, and G. Hinton, *Deep learning.* Nature, 2015. **521**(7553): p. 436-444.
4. Simonyan, K. and A. Zisserman, *Very deep convolutional networks for large-scale image recognition.* arXiv preprint arXiv:1409.1556, 2014.

513668 Immersive Introduction to Iot

Abstract:

ABSTRACT for AIChE 2018 Spring Meeting

Immersive Introduction to IoT

Joel Berg

The MathWorks, Inc.

joel.berg@mathworks.com

508-647-4415

You've heard of the "Internet of Things" (IoT), and have a notion of it, but what is it really? And how many presentations do you attend where the speaker actually asks you to turn on and use your phone?!

This talk will begin with a short overview of some common building blocks in IoT architecture, including sensing and measurement devices, wireless communications, cloud-based data aggregation, real-time model updating, and decision/action rules taken by cyber or physical actors.

Now the fun starts, with the presentation becoming more of a group "event." We will describe a process-oriented scenario, and ask audience members with smartphones to participate as actors within a living, cyber-physical model. Dynamic information will be shown and continuously updated on the main screen, and participants will play active roles according to signals they are seeing displayed and a brief set of instructions, by way of their devices (smartphones). The aggregated inputs of the audience will govern the system state evolution, which will also be displayed. The event will end with a short discussion and inventory of observations and insights deriving from the demonstration.

The objectives of this talk/event are to educate attendees about IoT, introduce the ThingSpeak® platform, provoke discussions about IoT use cases, and, last but not least, have some fun.

512435 Comparing Several Methods for Product Properties Prediction

Abstract:

Comparing several methods for product properties prediction

Jean Jérôme Da Costa[1], *Benoit Celse*[1*], *Fabien Chainet*[1], *Marion Lacoue Nègre*[1], *Cyril Ruckebush*[2], *Didier Espinat*[1]

1. *IFP Energies Nouvelles, Rond-Point de l'Echangeur de Solaize, 69360 Solaize, France*
2. *LASIR CNRS, Université Lille Nord de France, Sciences et Technologies, Cité scientifique, bât C5, 59665 Villeneuve d'Ascq Cedex - France*

**E-mail: benoit.celse@ifpen.fr*

HIGHLIGHTS

- Comparison between linear regression models and interpolation methods on toy examples
- Interpolation methods are much more efficient especially for non linear data
- Comparison between kriging and spline interpolation
- Kriging models can provide a posteriori analysis of the results
- Application on cloud point and viscosity index prediction

INTRODUCTION

In order to respect specification, petroleum products properties prediction is an increasingly need for refining industry. Machine learning is then more and more used to develop predictive models from process and physicochemical analytical data. The most classical models are based on classical (multilinear and nonlinear) regression referring to some equations which describe physicochemical phenomena or derived from empirical observations (Riazi 2005). However, regression models require to identify the adapted analytical expression for the regression function. That may be very complex in some situations.

Many chemometrics applications in modeling petroleum products properties were also proposed using partial least squares (PLS) regression (Braga, Jez Willian B., Junior, Araci Araújo Dos Santos et Martins 2014). Although these models provide relatively good performances, they use spectroscopic data such as NMR (nuclear magnetic resonance) or NIR (Near Infrared spectroscopy). This type of data are not currently available for refiners.

Some authors (Piloto-Rodríguez et al. 2013) proposed to use artificial neural networks (ANNs) for modeling complex properties. The results are good. However, the optimization of ANN requires to perform very stringent statistical steps to insure the model robustness and a lot of effort to prevent from over-fitting. This might be very tricky.

The aim of this work is to compare interpolation and linear regression methods on toy and real examples (could point, viscosity index ...) in order to suggest the best methodology to use for product properties prediction.

MATERIALS & METHODS

Kriging is an interpolation method based on stochastic multiGaussian assumptions (Roustant, Ginsbourger et Deville 2012)). Its goal is to provide the best linear unbiased estimate (B.L.U.E.) and was first used in applied geostatistics for modeling properties related to natural resources. More recently, other applications of kriging were developed such that metamodeling, which is an increasingly need in computing sciences (DiceKriging, DiceOptim: Two R Packages for the Analysis of Computer Experiments by Kriging-Based Metamodeling and Optimization 2012). One major point of kriging is the ability to provide a measure of the prediction uncertainty that depends on the data configuration (Roustant, Ginsbourger et Deville 2012).

The aim of splines interpolation is to draw contour lines "as smooth as possible", that is a map which looks like what a draftsman would obtain manually (Rooij, P. L. J. van et Schurer 1973). The splines consist of polynomials of degree m being local. The polynomials describe pieces of a line or surface. For degree m=1,2 or 3, a spline is called linear, quadratic or cubic respectively. This method is currently used in non-parametric statistical learning (Schumaker 2015).

RESULTS & DISCUSSIONS

Results on toy examples point out that interpolation methods have a high capacity of adaptability to linear or nonlinear situations without requiring previous analysis. Although, kriging and splines provide very close performance in some situations, kriging is more valuable since it provides a measure of uncertainties related to the predicted value based on stochastic assumptions. It is based on a specific distance between the new sample to predict and the calibration data base. This point is very important in modeling physicochemical properties of petroleum products. It is even particularly essential when the number of descriptors increases significantly (higher than 3).

Kriging and MLR models are developed and compared in real examples: cloud points and viscosity index. Although MLR provides good performances that are in accordance with the uncertainties of the standard measure, kriging enables to improve accuracy. In our knowledge, this is the first kriging model for the prediction of these two properties.

Globally, interpolation methods are well adapted to the modeling of petroleum products properties since they provide good performances when the number of samples is limited. Use of kriging methods to high dimensional study remains always challenging.

Bibliography

Braga, Jez Willian B.; Junior, Araci Araújo Dos Santos; Martins, Ingrid S. (2014) Determination of viscosity index in lubricant oils by infrared spectroscopy and PLSR. In : Fuel, vol. 120, p. 171–178. DOI: 10.1016/j.fuel.2013.12.017.

DiceKriging, DiceOptim: Two R Packages for the Analysis of Computer Experiments by Kriging-Based Metamodeling and Optimization (2012) : University of California, Los Angeles.

Piloto-Rodríguez, Ramón; Sánchez-Borroto, Yisel; Lapuerta, Magin; Goyos-Pérez, Leonardo; Verhelst, Sebastian (2013) Prediction of the cetane number of biodiesel using artificial neural networks and multiple linear regression. In : Global Conference on Renewable energy and Energy Efficiency for Desert Regions 2011 "GCREEDER 2011", vol. 65, n° Supplement C, p. 255–261. DOI: 10.1016/j.enconman.2012.07.023.

Riazi, M. R. (2005) Characterization and properties of petroleum fractions. 1st ed (Online-Ausg.). West Conshohocken, Pa : ASTM International (ASTM manual series MNL 50).

Rooij, P. L. J. van; Schurer, F. (1973) A bibliography on spline functions. Eindhoven : Technische Hogeschool (TH report / Technische Hogeschool Eindhoven Nederland, 73-WSK-01).

Roustant, Olivier; Ginsbourger, David; Deville, Yves (2012) DiceKriging, DiceOptim: Two R Packages for the Analysis of Computer Experiments by Kriging-Based Metamodeling and Optimization. In : Journal of statstical software, vol. 51, n° 1.

Schumaker, Larry L. (2015) Spline functions. Computational methods. Philadelphia : Society for Industrial and Applied Mathematics (Other titles in applied mathematics, 142).

512697 Big Data to Support Operators in Chemical Plants

Abstract:

Big Data to support operators in chemical plants

Dr. Zied Ouertani, Dr. Benjamin Klöpper

ABB

Abstract

Big data technologies enable new possibilities to analyze historical data generated by process plants. This contribution provides an overview of how big data technology is applied to analyze rare events and hence support maintenance operations by detecting anomalies that would have gone otherwise unnoticed

Introduction –Big Data for Chemicals industry

A typical chemical plant generates large amounts of data throughout it whole life cycle: I/O and tag lists, piping and instrumentation diagrams (P&ID), control logic, alarm configurations (during planning and commissioning), measurement values, alarm and event logs, shift books, laboratory results (during operation), maintenance notification, repair and inspection reports (during maintenance). Today, the scope of analytics is usually limited to a single data source. One common example of such an analysis is loop performance monitoring: Evaluating the control quality of single control loops based on measurements, set point, and controller output [1]. However, the limitation to a single data sources hampers the process of learning from historical data. An integration of all different types of data within process plants will bring up the notorious big data attributes: high volume, high velocity, and high variety [1]. For instance, a refinery can generate more than 300 GB measured values per year, produced by more than 60,000 sensors with sampling rates between 1 and 60 seconds. Data exists in structured (sensor readings, database tables), semi-structured (alarm and event logs) and unstructured formats (shift books, operation manuals). Often, data has been often stored for 10 years and even longer.

This contribution explores the potential of Big Data Analytics in the context of chemical plants with a particular focus on operator support. It focuses on specific application scenarios of event prediction and anomaly detection.

Big Data application addressing rare events for anomaly detection

Monitoring a chemical plant is certainly a demanding and cognitive complex task. More than thousands signals with trend displays and faceplates distributed over dozens of operator screens are not a rarity. In calm situations without alarms or specific tasks (e.g. ramp-up, load change, shutdown) operators screen the distributed control system for abnormalities or suspicious signals. Operators perform this task by browsing through operator screens and trend displays in a more or less systematic fashion. This unstructured procedure obviously heavily depends on the experience level of the operator and carries a high risk to miss relevant abnormalities. Especially for unexperienced operators it is difficult to judge if the trend of a specific signal actually is abnormal or not.

A big data support system should address these issues, direct the operator's attention to relevant signals, and support the diagnosis and assessment of the abnormality. This implies a quick visual impression off the plant status and support for putting suspicious signals into context and predict unexpected behavior.

Based on a distance based anomaly detection algorithms, a Big Data Anomaly detection is developed for the processing framework Apache Spark. Apache Spark enables the algorithms to scale to large data sets by adding additional computational resources to the systems (scale out). REF _Ref474483343 \h Figure 1 08D0C9EA79F9BACE118C8200AA004BA90B02000000080000000E0000005F0052006500660034003700340034003800330033003400330000000 one shows the big data architecture of the plant-wide anomaly detection. The architecture is an adaption of Maerz Lamdba-Architecture [3] and can be organized in three different layers: stream layer (handling and processing streams of incoming data), batch layer (processing large amounts of stored historical data), and the serving layer (providing access to the algorithm results).

The ingestion point for new data is Apache Kafka, a stream-processing framework, high-throughput, low-latency platform for handling real-time data feeds. Its capability to persist data for long time and leave control when to read data to the consumer of data makes it ideal to decouple the fast reading process of the stream analytics and the more batch and bulk oriented reading process to the storage of the batch layer.

Figure SEQ Figure * ARABIC 1: Big Data System Architecture for Plant-Wide Anomaly-Detection

The batch layer consists of two elements: the Hadoop File Systems (HDFS) that stores process data in columnar fashion and enables easy distributed processing by the batch algorithm executed within Apache Spark. The implemented k-nearest neighbor algorithm analyzes the proximity between episodes (in our case 60 minutes) of data from single signals (univariate approach). In order to handle the large amount of data without extensive expert classification by into normal and abnormal behavior, the big data algorithm works in an unsupervised fashion considering all available data (normal and abnormal) and applying the 'three-sigma rule of thumb' to initial identify and score anomalies. The results of the algorithm are the anomaly scores for the historical episodes. The results and the historical episodes are stored in the serving layer database (HBASE).

The second part of the streaming layer is Apache Spark Streaming. Spark Streaming is a library for Spark and represents the possibility to use Spark over a stream of data. Spark Streaming uses a micro-batch architecture, in the sense that the streaming computation is treated as a continuous series of batch computations on small batches of data. The duration of a batch is determined by the configuration parameter batch interval. When a batch starts, all the data that arrive during the interval are added to the batch. This micro-batching semantics fits the anomaly detection scenario very well and enables to share the algorithm code between the batch and streaming layer. While the batch algorithm compares all historical episodes with all other historical episodes and determines a proximity threshold, the stream algorithm compares one query episode against all historical episode and calculates an anomaly score based on the threshold value. The results of the stream algorithms are stored in the serving layer database Apache HBASE.

HBASE is a distributed key-value storage running on top of HDFS. HBASE is very scalable, robust and has sufficient answering times for the anomaly detection scenario. The operator UI, which is, deploy via an application server polls the HBASE for the relevant data. REF _Ref474499904 \h Figure 2 08D0C9EA79F9BACE118C8200AA004BA90B02000000080000000E0000005F005200650066003400370034003400390039003900300034000000 shows a screenshot of the operator UI. The heat map (1) in the upper left corner shows the most suspicious signals according to the anomaly score calculated by the streaming algorithm. The heat map also gives an indication of the development of the anomaly score over time. The upper right corner (2) shows a normal episode (low anomaly score), the lower left hand corner (3) the current trend of signal and the right hand corner (4) a similar situation in the past. This information helps the operator to perform a broad monitoring of the plant and to judge if a signal trend is actually abnormal or not. The similar situation in the past helps to diagnose the current situation and can be used to investigate additional data sources like shift books or alarm logs.

Figure SEQ Figure * ARABIC 2: Anomaly Detection UI

ACKNOWLEDGMENTS

The authors would like to thank the development partners from the KDE Group, University of Kassel, the Department of Measurement and Control, University of Kassel, the Chair of Process Control System Engineering, TU Dresden and RapidMiner for the great cooperation and joint work in the project. The development team is thankful to PCK, BASF and INEOS for many valuable discussions, crucial feedback and for providing access to the data.

Contact:

Dr. Zied Ouertani
phone: +49 (0) 6213 811012
e-mail: mohamed-zied.ouertani@de.abb.com

Dr. Benjamin Klöpper
phone: +49 (0) 6203 716211
e-mail: benjamin.kloepper@de.abb.com

References

[1] Atzmueller, Martin; Kloepper, Benjamin ; Mawla, Hassan Al ; Jäschke, Benjamin ; Hollender, Martin ; Graube, Markus ; Arnu, David ; Schmidt, Andreas ; Heinze, Sebastian ; Schorer, Lukas ; Kroll, Andreas ; Stumme, Gerd ; Urbas, Leon: {Big Data Analytics for Proactive Industrial Decision Support: Approaches \& First Experiences in the Context of the FEE Project}. In: atp edition, 58 (2016), Nr. 9

[2] McMillan, Gregory K. (2014) Tuning and control loop performance.Momentum Press

[3] McAfee, A., Brynjolfsson, E., Davenport, T.H., Patil, D.J. Barton, D. (2012) Big Data. The management Revolution. Harvard Bus Rev 90(10), 61-67

512519 Finding a Needle in an Ocean of Data with Digital Twins: An Oil & Gas Perspective on Analytics

Abstract:

Analyzing industrial data requires a fundamentally different approach than consumer data. This is due to a confluence of several factors. Compared to traditional ERP systems that store transactional data like logistics reports, maintenance logs, inventory logs, etc., Oil & Gas industrial data is 10-100X more voluminous, is collected at 100-1000X more frequently and is 5-10X more diverse. The hybrid modeling framework to build Digital Twins that combines physics-based models with deep learning techniques will be highlighted with specific application to system level optimization. We will present novel application of Digital Twins and deep learning models for anomaly detection in process models, network modeling and process optimization.

513738 Approaches to Multiblock Modeling and Dimension Reduction for Combining Disparate Data Types

Abstract:

Continuous manufacturing, which is being strongly promoted by the FDA, is an innovative approach to pharmaceutical processing that utilizes information from multiple sensors throughout the entire process. Due to the numerous factors that need to be monitored and controlled at the same time, a multi-step approach is necessary for optimizing individual unit operations as well as the overall process. In this presentation we will show some approaches to combining process and spectroscopic data, which can be applied to process monitoring for continuous manufacturing as well as for batch processes.

AUTHOR INDEX